MATHEMATICS RESEARCH DEVELOPMENTS

BOOLEAN FUNCTIONS

THEORY, FUNDAMENTALS AND ENGINEERING APPLICATIONS

MATHEMATICS RESEARCH DEVELOPMENTS

Additional books in this series can be found on Nova's website under the Series tab.

Additional e-books in this series can be found on Nova's website under the e-book tab.

MATHEMATICS RESEARCH DEVELOPMENTS

BOOLEAN FUNCTIONS

THEORY, FUNDAMENTALS AND ENGINEERING APPLICATIONS

ALLEN HINES
EDITOR

nova publishers
New York

Library of Congress Cataloging-in-Publication Data

Boolean functions : theory, fundamentals and engineering applications / Allen Hines, editor.
pages cm. -- (Mathematics research developments)
Includes bibliographical references and index.
ISBN 978-1-63483-023-2 (hardcover)
1. Algebra, Boolean. 2. Symmetric functions. I. Hines, Allen, editor.
QA10.3.B64 2015
511.3'24--dc23

2015019751

Published by Nova Science Publishers, Inc. † New York

CONTENTS

Preface		**vii**
Chapter 1	Symmetric Boolean Functions *Peter M. Maurer*	**1**
Chapter 2	Boolean Functions: All-Optical Implementation Using Quantum-Dot Semiconductor Optical Amplifiers in Mach-Zehnder Interferometer *K. E. Zoiros, E. Dimitriadou and T. Houbavlis*	**57**
Bibliography		**117**
Index		**131**

PREFACE

Symmetric Boolean functions have played an important role in many aspects of design automation for many years. This book summarizes developments and provides a collection of new tools and techniques that can be used to advance the study of Boolean functions. Moreover, Boolean functions provide the necessary framework for expressing the operation of logic gates, which are the key building units for the accomplishment of signal processing tasks in fundamental and system-oriented levels. The book concludes with a discussion on how Boolean functions can be used to ensure the minimum degree of logical functionality between light-wave modulated signals.

Chapter 1 – The purpose of this chapter is to summarize what is known about symmetric Boolean functions, but will concentrate primarily on my own work. I will discuss several new types of symmetry, along with tools and analytical techniques for detecting and studying such symmetries. First, I will provide a comprehensive introduction to the mathematics of group theory, and show how this serves as a basis for the study of symmetric Boolean functions. I will extend these ideas to matrix-based symmetry, and show how matrix-based symmetry greatly expands the concept of permutation-based symmetry. The new types of symmetry discussed in this chapter will include conjugate symmetry, super-symmetry, exotic symmetry, and hyper-symmetry. The names "super-symmetry," "exotic symmetry," and "hyper symmetry" have not appeared in print and will be defined in the chapter. These are all matrix-based symmetries. In addition I will discuss the existing symmetry types: anti-symmetry, hierarchical symmetry, and Kronecker symmetry and potential extensions to such.

Chapter 2 – Boolean functions provide the necessary framework for expressing the operation of logic gates, which are the key building units for the accomplishment of signal processing tasks in fundamental and system-oriented level. With the unceasing increase of digital information volume, the capability of executing these functions has been progressively evolved from electronics to photonics so as to avoid the severe constraints imposed in the former domain. This book chapter reviews how this can be done exclusively by means of light by exploring the novel technology of quantum-dot semiconductor optical amplifiers (QD-SOAs), which are incorporated in a switching structure formed by the Mach-Zehnder interferometer (MZI). For this purpose we initially present some state-of-the-art technologies that are available for realizing all-optically logic operations, with particular emphasis on the well-established option of SOAs. We underline the limitations in handling ultrafast information when using conventional-type SOAs and we comment on the various solutions that have been proposed for overcoming them. We highlight the need for adopting the novel approach of QD-SOAs in order to successfully confront the challenge of keeping pace with the rapid increase of single channel data rates. We justify this choice by referring to the attractive inherent properties of these devices and how they are combined with those of the MZI to achieve ultra high speed switching. We physically and theoretically describe the operation of QD-SOAs, first as stand-alone entities and then when they are incorporated in the MZI. We demonstrate how the basic configuration of the latter can be modified and driven so as to produce different Boolean functions at its two output ports in a versatile and reconfigurable manner. We identify a set of such functions and explain the reasons for their selection according to their role and significance. We define performance criteria and adapt them to the special nature of the logical outcome of each target function. We provide simulation results concerning the impact of critical operating parameters on these metrics and based on their concise interpretation we give guidelines for the execution of each Boolean function both with logical correctness and high quality. We also address several practical issues regarding the real implementation of these functions. Finally, we report on characteristic lightwave digital applications of enhanced functionality enabled through complex circuits, which are synthesized and formulated by combinations of the considered QD-SOA-based MZI Boolean functions.

In: Boolean Functions
Editor: Allen Hines
ISBN: 978-1-63483-023-2

Chapter 1

SYMMETRIC BOOLEAN FUNCTIONS

Peter M. Maurer
Dept. of Computer Science, Baylor University,
Waco, Texas, US

ABSTRACT

The purpose of this chapter is to summarize what is known about symmetric Boolean functions, but will concentrate primarily on my own work. I will discuss several new types of symmetry, along with tools and analytical techniques for detecting and studying such symmetries. First, I will provide a comprehensive introduction to the mathematics of group theory, and show how this serves as a basis for the study of symmetric Boolean functions. I will extend these ideas to matrix-based symmetry, and show how matrix-based symmetry greatly expands the concept of permutation-based symmetry.

The new types of symmetry discussed in this chapter will include conjugate symmetry, super-symmetry, exotic symmetry, and hyper-symmetry. The names "super-symmetry," "exotic symmetry," and "hyper symmetry" have not appeared in print and will be defined in the chapter. These are all matrix-based symmetries. In addition I will discuss the existing symmetry types: anti-symmetry, hierarchical symmetry, and Kronecker symmetry and potential extensions to such.

1. Introduction

Symmetric Boolean functions have played an important role in many aspects of design automation for many years. In fact, the study of symmetric Boolean functions predates the development of design automation as a distinct science by several years.

The purpose of this chapter is to summarize developments so far and to provide a collection of new tools and techniques that can be used to advance the study of such functions. I will show how these tools and techniques are natural extensions of existing techniques that have been used in the design automation community and in the mathematical study of symmetry.

First, I will provide a comprehensive introduction to the mathematics of group theory, and show how this serves as a basis for the study of symmetric Boolean functions. In particular I will discuss *conjugate subgroups*, *forbidden subgroups*, and *k-transitivity*. These are important topics that, as far as I know, have not been discussed in the context of design automation.

I will introduce a new concept called *Boolean orbits* that can be used to characterize any known form of symmetry. I will use the concept of Boolean orbits to create a symmetry detection algorithm that is capable of detecting any known form of symmetry, including the new types of symmetry introduced in this paper.

I will provide an exhaustive list of the positive ordinary symmetries of Boolean functions of six or fewer inputs. The purpose of this discussion is to demonstrate increasing complexity of symmetry types as the number of function inputs increases. This work suggests that no matter how many different types of symmetry that have been identified that there will always be new types of symmetry introduced as the number of function inputs increases.

By far the most important contribution of this chapter is the matrix approach to symmetry. Everything that can be done using permutations can be translated, quite trivially, into the matrix domain. However, the matrix domain provides many new classes of symmetry that are virtually inaccessible by any other method. Some of these symmetries have been explored in a limited way using other techniques, but the matrix domain allows these symmetries to be explored much more thorough way than is possible using other techniques. In particular, the matrix domain permits the exploration of *conjugate symmetries* and *super symmetries*. In addition, the matrix approach permits the definition of two entirely new types of symmetry known as *exotic symmetry* and *hyper symmetry*.

Using the tensor product and direct products of matrices it is possible to give a more extensive definition of hierarchical symmetry, and to demonstrate how various types of independent and lock-step sub-symmetries arise.

The earliest work on symmetric Boolean functions, with respect to circuit design, was done by Shannon [1]. This work was primarily concerned with total and partial symmetries of functions with multi-phase inputs. The phase of an input indicates whether it is active-high or active-low. Shannon was primarily interested in symmetry as an aid to synthesizing relay circuits. Relays can be constructed as either "normally open" or "normally closed" with no cost penalty which allows active-high and active-low inputs to be mixed arbitrarily. In the VLSI arena enhancement-mode and depletion-mode transistors are the direct analog of normally-open and normally-closed relays. Because modern VLSI technology makes little or no use of depletion-mode transistors, it may seem safe to ignore functions with multi-phase inputs. This is not the case, however, and virtually all recent work in symmetric Boolean functions includes the multi-phase case.

2. The Permutation Group of a Function

2.1. Introduction

Informally, a symmetric Boolean function in n variables is a function whose variables can be rearranged in some fashion without changing the output of the function. For example, consider the 3-input function $f = x_1x_2x_3$. (In this chapter, Boolean functions will be represented as expressions using the operators AND, OR and NOT. AND will be represented as multiplication, OR will be represented as addition, and NOT will be represented using the prime symbol. Thus $x_1x_2' + x_3$ represents the function $(x_1\ AND\ (NOT\ x_2))\ OR\ x_3$.) The function $f = x_1x_2x_3$ is identical to the function $g = x_2x_1x_3$ and the function $h = x_3x_2x_1$. In fact the three variables of f can be rearranged arbitrarily without changing the value of its output. Other functions, such as $k = x_1x_2 + x_3$, have more restricted symmetry. It is possible to swap the variables x_1 and x_2 of k, but x_3 cannot be swapped with x_1 or x_2.

Symmetries such as those exhibited by f and k can be dealt with informally. The function f exhibits what is normally called *total symmetry*. Total symmetry permits all variables of a function to be rearranged arbitrarily. The function k exhibits *partial symmetry*. Partial symmetry permits one or more subsets of variables to be rearranged arbitrarily, but the rearrangements are restricted to these subsets.

More complex types of symmetry require a more systematic approach.

2.2. Permutations

Let X be any finite set. A *permutation* of X is a one-to-one onto function from X to itself. (A function, p, is one-to-one if $p(a)=p(b)$ implies that $a=b$. It is onto if for every element, b, of X, there is an $a \in X$ such that $p(a)=b$.) The concept of a permutation formalizes the idea of rearranging a set of objects. Suppose X is the finite set $\{1,2,3\}$, and the function p is defined as follows: $(p(1)=2, p(2)=3, p(3)=1)$. One can envision p as a rearrangement of the elements of X. The element 2 takes the place of 1, 3 takes the place of 2, and 1 takes the place of 3. Every one-to-one onto function on X corresponds to a rearrangement of X, and every rearrangement of X corresponds to a one-to-one onto function on X.

The set of all permutations of X is called the *symmetric group* on X, and is written S_X. When X is the set $\{1,2,3,...,n\}$ the symmetric group on X is written S_n. If two sets, X and Y are the same size, then S_X is identical (i.e. isomorphic) to S_Y. Because of this, it is better to concentrate on S_n, and assume that the set X is indexed by the elements of the set $\{1,2,3,...,n\}$, where n is the size of X. An obvious example is our manner of writing input variables as $\{x_1, x_2, x_3, ..., x_n\}$. The size of S_n is $n!$.

Permutations can be written in several different ways. The most straightforward way is to list the elements of $\{1,2,3,...,n\}$ on one row, and the corresponding values of p on the second row, as in Figure 1.

$$r = \begin{pmatrix} 1 & 2 & 3 & 4 \\ 2 & 3 & 4 & 1 \end{pmatrix}$$

Figure 1. Double List Notation.

Of course, one could observe that the top row is identical for all permutations of $\{1,2,3,4\}$ and shorten the notation to $p = \begin{pmatrix} 2 & 3 & 4 & 1 \end{pmatrix}$, or even 2,3,4,1.

A more efficient way to represent permutations is cycle notation. A cycle is a set of elements from $\{1,2,3,...,n\}$ that are shifted cyclically by the permutation. Every permutation can be written as one or more cyclic shifts of elements. A cyclic shift is written as a list of elements that are shifted in the order in which they are listed. The entire list is enclosed in parentheses.

A cycle that shifts n elements is called an *n*-cycle. The simplest cyclic shift is the swapping of two elements, which is also called a *transposition*, or a 2-cycle.For example consider the permutation s of Figure 2.

$$s = \begin{pmatrix} 1 & 2 & 3 & 4 \\ 2 & 1 & 3 & 4 \end{pmatrix}$$

Figure 2. Swapping 1 and 2.

In cycle notation s would be written $(1,2)$ because zero is moved to position one and one is moved to position zero. The elements 3 and 4 are not moved and are omitted. In cycle notation, the permutation r of Figure 1 would be written (1,2,3,4). This notation implies that one replaces two, two replaces three, three replaces four, and four replaces one. Because the shift is cyclic, the permutation could be written in four ways: (1,2,3,4), (2,3,4,1), (3,4,1,2) and (4,1,2,3). These four cycles are identical, but to avoid confusion the usual convention is to write the smallest element of the cycle first. This gives a unique representation for any cyclic shift. Some permutations contain more than one cyclic shift. For example, consider permutation t of Figure 3.

$$t = \begin{pmatrix} 1 & 2 & 3 & 4 \\ 2 & 1 & 4 & 3 \end{pmatrix}$$

Figure 3. A double cycle.

The permutation t contains two distinct cyclic shifts, 1 is swapped with 2 and 3 is swapped with 4. In cycle notation this is written $(1,2)(3,4)$. This type of notation is called a *product* of cycles and signifies that one cyclic shift must be followed by another. The two cycles $(1,2)$, and $(3,4)$ are called *disjoint*, because they don't share any elements. If two cycles are disjoint, it doesn't matter in which order they are written, so $(3,4)(1,2)$ is the same as $(1,2)(3,4)$. However, the usual convention is to sort a product of disjoint cycles into ascending order first by the length of the cycle and then by the first element of the cycle. Every permutation, except the identity permutation, can be written as a product of disjoint cycles. The identity permutation leaves everything fixed and is written I.

The *degree* of a permutation is the size of the underlying set, so every element of S_n is of degree *n*. Although cycle notation has a number of advantages, it does not allow us to determine the degree of a permutation. For example, the permutation (1,2,3) could be a degree-3 permutation on the set {1,2,3} or a degree-4 permutation on the set {1,2,3,4}. This information must be supplied by context.

Permutations can be "multiplied" by performing one after the other. This is actually function composition, but it's easier to focus on the idea of performing permutations sequentially. For example, let us multiply the permutation $(1,2,3,4)$ by itself. Figure 4 shows the result of this multiplication.

$$(1,2,3,4)(1,2,3,4)=(1,3)(2,4)$$

Figure 4. Multiplying permutations.

When multiplying two permutations, it is necessary to determine what happens to each element. The process is shown step-by-step in Figure 5.

1. Start with the smallest element, 1.
2. Observe that Element 1 is mapped to 2 by the first permutation and element 2 is mapped to 3 by the second permutation, so the product maps element 1 to element 3.
3. Continue with the next element of the permutation being created, element 3. Element 3 is mapped to element 4 by the first permutation,

and element 4 is mapped to element 1 by the second, so element 3 is mapped to element 1 by the product. This completes the cycle.

4. Proceed by selecting the smallest unprocessed element and starting a new cycle. The smallest unprocessed element is element 2.
5. Observe that the element 2 is mapped to element 3 by the first permutation and element 3 is mapped to element 4 by the second.
6. Observe that the element 4 is mapped to element 1 by the first permutation and element 1 is mapped to element 2 by the second permutation. This completes the second cycle. Since the product permutation now contains all elements of {1,2,3,4} and is complete.

Step 1

$$(1,2,3,4)(1,2,3,4)=(1,$$

Step 2:

$$(1,2,3,4)(1,2,3,4)=(1,3$$

Step 3:

$$(1,2,3,4)(1,2,3,4)=(1,3)$$

Step 4:

$$(1,2,3,4)(1,2,3,4)=(1,3)(2,$$

Step 5:

$$(1,2,3,4)(1,2,3,4)=(1,3)(2,4$$

Step 6:

$$(1,2,3,4)(1,2,3,4)=(1,3)(2,4)$$

Figure 5. Step-by-step multiplication.

Sometimes cycle notation doesn't list every element of the underlying set. When this happens, it implies that the permutation maps the element to itself. The multiplication procedure will often give result in a "cycle" with a single element, such as (2). These are eliminated from the final cycle representation. Figure 6 gives two products where these rules must be applied.

(1,2,3)(2,3,4)=(1,3)(2,4)
(2,3)(1,2,3,4)=(1,2,4)(3)=(1,2,4)

Figure 6. Multiplication problems.

2.3. Permutation Groups

A *group* is a set, G, with a multiplication operator, $\circ$, which has the following four properties.

1. For any two (not necessarily distinct) elements $a,b \in G$, $a \circ b$ is defined and is an element of G.
2. For any three elements $a,b,c \in G$, $(a \circ b) \circ c = a \circ (b \circ c)$.
3. There exists an $i \in G$, called the *identity* element of G, such that $i \circ a = a \circ i = a$ for all $a \in G$.
4. For all $a \in G$ there exists $a^{-1} \in G$, called the *inverse* of a, such that $a \circ a^{-1} = a^{-1} \circ a = i$.

The set of all permutations on a set of size *n*, S_n, is a permutation group. Many subsets of S_n also are groups. A set $H \subseteq G$ is a *subgroup* of G, if H, itself, has the four above properties. Since G is assumed to be finite, one can show that any subset of G that has the property 1 also has properties 2, 3, and 4. Thus it is sufficient to test whether a subset of G has property 1.

At times it is inconvenient to list all elements of a group. In most cases it is possible to list only a few elements and then use the principle that a group is closed under multiplication to imply the rest. For example, every element of S_3 can be written as a product of the cycles (0,1) and (0,1,2), as Figure 7 shows.

We say that S_3 is *generated* by the set $\{(0,1),(0,1,2)\}$. Another way to write this is $S_3 = \langle (0,1),(0,1,2) \rangle$. Sometimes it is more convenient to list the generators of a group than to list every element of the group, although looking at the generators of a group doesn't always reveal any details about its structure.

$$(0,1)(0,1) = I$$
$$(0,1) = (0,1)$$
$$(0,1)(0,1,2) = (0,2)$$
$$(0,1,2)(0,1) = (1,2)$$
$$(0,1,2) = (0,1,2)$$
$$(0,1,2)(0,1,2) = (0,2,1)$$

Figure 7. The generators of S_3 .

A similar concept is that every *n*-cycle can be written as a product of 2-cycles. Naturally, these 2-cycles will *not* be disjoint. For example, the cycle (0,1,2,3) can be written as (0,1)(0,2)(0,3). If we multiply this out, we will obtain the original permutation. This decomposition of an *n*-cycle is not unique, nor is the number of 2-cycles fixed, since (0,1)(0,1)(0,1)(0,2)(0,3)=(1,2)(1,3)(1,0)=(0,1,2,3). However, the *parity* of the decomposition is fixed. If a permutation can be decomposed into an odd number of 2-cycles, then there is no way to decompose it into an even number of 2-cycles, and vice-versa. This allows us to characterize permutations as even or odd, depending on the number of 2-cycles in the decomposition. Since every *n*-cycle can be decomposed into $n-1$ 2-cycles, the parity of a permutation is readily apparent from the cycle notation. The distinction between even and odd permutations will be useful later in this chapter.

New subgroups can be created by finding intersection and union of two subgroups. Suppose H and K are two subgroups of a group G . The intersection $H \cap K$ will always be a subgroup of G and will never be empty because I must be an element of every subgroup. However the union $H \cup K$ often fails to be a subgroup, because it won't be closed under multiplication. For example, consider the two subgroups of S_3 , $\{I,(0,1)\}$ and $\{I,(0,1,2),(0,2,1)\}$. The union of these is $\{I,(0,1),(0,1,2),(0,2,1)\}$, but this set is not closed under multiplication because $(0,1)(0,1,2) = (0,2)$. If it is necessary to discuss the union of two groups, we generally treat $H \cup K$ as a set of generators and talk about the group generated by $H \cup K$, $\langle H \cup K \rangle$.

2.4. Symmetric Functions

All Boolean functions are invariant with respect to the identity permutation, but the *symmetric* Boolean functions are invariant with respect to other permutations as well. Definition 1 formalizes this concept.

Definition 1. Let f be an n-input Boolean function. If there is a permutation $p \neq I$ such that $p(f) = f$ then f is said to be *symmetric*.

Given an n-input Boolean function f, let H be the set of all permutations that leave f invariant. As noted above, H must be a subgroup of S_n. We want to capture this concept in a formal definition, Definition 2.

Definition 2. Let f be an n-input Boolean function and let H be the group of degree n permutations that leave f invariant. Then H is called the *symmetry group* or *symmetry rule* of f.

The symmetry of a Boolean function is completely characterized by its symmetry group. The symmetry group always exists, even if the function is not symmetric, because every Boolean function is invariant with respect to the identity permutation. A non-symmetric function has the symmetry group $\{I\}$.

The symmetry groups of some functions are easy to determine. The symmetry group of the three-input function $x_0x_1x_2$ is $S_3 = \{I,(0,1),(0,2),(1,2),(0,1,2),(0,2,1)\}$, while the symmetry group of the four-input function $x_0 + x_1x_2x_3$ is $\{I,(1,2),(1,3),(2,3),(1,2,3),(1,3,2)\}$. Our intuition tells us that these two groups should be the same, even though the permutations are different. Our intuition is correct if we make the correspondence shown in Figure 8.

$$
\begin{array}{c}
I \text{ -- } I \\
(0,1) - (1,2) \\
(0,2) - (1,3) \\
(1,2) - (2,3) \\
(0,1,2) - (1,2,3) \\
(0,2,1) - (1,3,2)
\end{array}
$$

Figure 8. A correspondence between groups.

It's clear that the correspondence shown in Figure 8 is "correct" because the corresponding elements "multiply the same way." Some other correspondence, such as one that associates (0,1) with (1,2,3) would be "incorrect" because $(0,1)^2 = (0,1)(0,1) = I$, and $(1,2,3)^2 = (1,2,3)(1,2,3) = (1,3,2)$.

Our intuitive idea of "correctness" corresponds to the formal concept of a homomorphism. Definition 3 gives the formal definition of a homomorphism.

Definition 3. Let G and H be groups and let F be a function that maps the elements of G into the elements of H. F is a *homomorphism* if for all $a,b \in G$, $F(ab) = F(a)F(b)$. If the function F is one-to-one, then F is called an *isomorphism.* If there is an isomorphism from G to H and an isomorphism from H to G, then G and H are said to be *isomorphic.*

If two groups are isomorphic, then they are the same group with different names for the elements. For two groups to be isomorphic, they must be the same size. This is why Definition 3 states that there must be two isomorphisms, one in each direction. There is an isomorphism from S_3 to S_4 (Figure 8, for example) but there is no isomorphism from S_4 to S_3.

Isomorphism is a powerful idea, but in some cases it isn't enough to distinguish between symmetry groups that are intuitively different from one another. For example, consider the two isomorphic subgroups of S_4, $\{I,(0,1)\}$ and $\{I,(0,1)(2,3)\}$. In the first group allows us to swap inputs 0 and 1, while the second says that it's OK to swap inputs 0 and 1, but if we do so we must also swap inputs 2 and 3. Clearly these symmetry rules are different even though the groups are isomorphic. To distinguish between these two groups, we need the concept of *conjugacy.* This concept is straightforward, but the mathematical definition can be confusing. Intuitively, two symmetry groups are conjugate if they "do the same thing" but to different sets of inputs.

To illustrate the concept of conjugacy, let us assume that we have a board with five disks as illustrated in Figure 9. As the board moves down the conveyor belt, it passes through a machine that drills a hole through the center disk. However, we don't want to drill the center disk, we actually want to drill the second disk from the left. It's too expensive to move the drill, but it's very inexpensive to install machinery to move the disks around. We can drill the correct disk by using the system pictured in Figure 10.

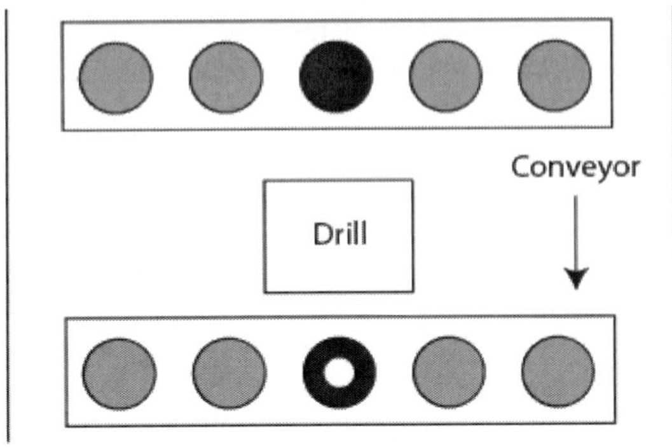

Figure 9. Drilling a hole in a disk.

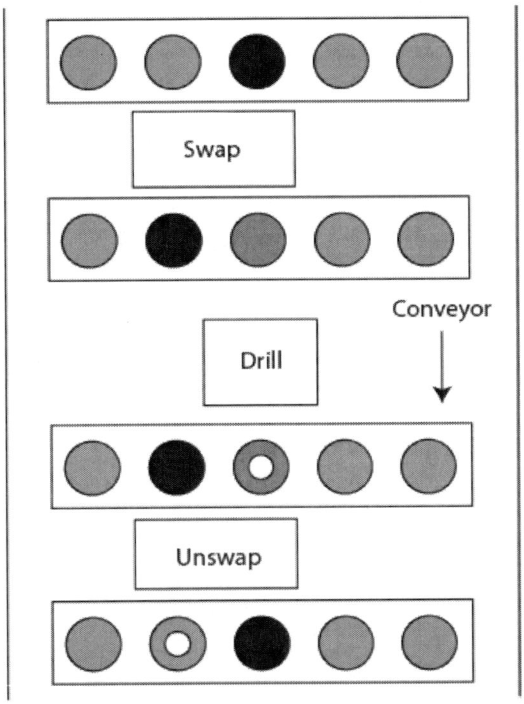

Figure 10. Drilling a different disk.

Figures 9 and 10 show two drilling operations that are *conjugate* to one another. Drilling a hole in disk 2 is conjugate to drilling a hole in disk 3 because we can rearrange the disks so that disk 2 is in the proper position for drilling and then put the disks back the way they were.

Conjugate elements in permutation groups are based on the same idea. Swapping inputs 2 and 3 of a 4-input function is conjugate to swapping inputs

0 and 1, because we can move inputs 3 and 4 into positions 0 and 1, swap them, and then move them back into positions 3 and 4. This shows that the permutation (0,1) is conjugate to the permutation (2,3). On the other hand the permutation (0,1)(2,3) cannot possibly be conjugate to (0,1), because no matter how we rearrange the inputs of a function, the permutation (0,1) swaps one, and only one pair of inputs.

This concept tends to be a little confusing because the operation that we wish to perform is a permutation, and the mechanism that we use to move elements into the proper position is also a permutation. It is helpful to remember that we are viewing permutations in two very different ways when we deal with conjugates. The formal definition of conjugacy is given in Definition 4.

Definition 4. Let p and q be two elements of G. The elements p and q are said to be *conjugates* if there is an $r \in G$ such that $rpr^{-1} = q$. Let H and K be subgroups of G. H and K are said to be *conjugates* if there exists an $r \in G$ such that $rHr^{-1} = K$, where $rHr^{-1} = \left\{ rpr^{-1} \mid p \in H \right\}$.

The expression $rpr^{-1} = q$ means that we could either use q directly, or we could first use r to rearrange some inputs, then apply p to perform the desired transformation, and then use r^{-1} to put things back in their starting positions. We would never actually perform the sequence of operations rpr^{-1}. This is merely a mechanism for showing that p and q do the same thing, but to different sets of inputs.

We can find the conjugate of a subgroup by finding the conjugates of every element of the subgroup. If two subgroups are conjugate to one another, they are also isomorphic, but two isomorphic subgroups are not necessarily conjugate to one another.

The identity element has no conjugate elements other than itself since $rIr^{-1} = rr^{-1} = I$ for any r. To compute conjugates we must be able to compute the inverse, r^{-1}, of a permutation r. Cycle notation makes it easy to do this. Any two-cycle is the inverse of itself. For any other n-cycle, leave the first element unchanged, and write the remaining elements in reverse order. For a product of disjoint cycles, find the inverse of each cycle. Thus the inverse of (0,1,2)(3,4,5,6)(7,8) would be (0,2,1)(3,6,5,4)(7,8). Figure 11 shows the conjugate elements of S_3 and how they are found.

$$\begin{array}{ll} & \{I\} \\ (0,2)(0,1)(0,2) = (1,2) & \{(0,1),(1,2),(0,2)\} \\ (1,2)(0,1)(1,2) = (0,2) & \\ (1,2)(0,1,2)(1,2) = (0,2,1) & \{(0,1,2),(0,2,1)\} \end{array}$$

Figure 11. The conjugates of S_3 .

Two permutations are conjugate to one another if and only if they are of the same degree and have the same cycle structure. This is not true for subgroups, however. Figure 12 shows two subgroups of S_6 that have identical cycle structure, but are not conjugate to one another.

$$\{I,\ (2,3)(4,5),(0,1)(4,5),\ (0,1)(2,3)\}$$

$$\{I,\ (0,1)(2,3),(0,2)(1,3),\ (0,3)(1,2)\}$$

Figure 12. Non-conjugate groups with identical cycle structure.

3. Categorizing Symmetries

3.1. Introduction

There are many recognized types of symmetry, each one of which can be associated with certain types of subgroups of S_n . Unfortunately, a systematic characterization of *all* types of symmetry seems to be a difficult problem. For small n , one can simply enumerate all subgroups of S_n , but this approach quickly hits a computational barrier. One problem is the size of S_n which grows factorially with n . Another problem is that S_n contains n isomorphic copies of S_{n-1}. If we eliminate a single input and enumerate all the permutations that affect only the remaining inputs, this set of permutations will be isomorphic to S_{n-1}.

There are n different variables that can be omitted, giving n copies of S_{n-1}. In addition, there are subgroups that affect all of the inputs so the number of subgroups grows more than factorially with respect to n. In addition, as n increases, it becomes increasingly difficult to handle the individual permutations.

We have been able to enumerate the subgroups of S_2 through S_8 using ordinary desk-top computers (and some clever algorithms).Using better algorithms it has been possible to enumerate the subgroups of all symmetric groups up to and including S_{18} [2]. Even these more sophisticated algorithms will probably not work much beyond S_{20}.

3.2. Basic Symmetry Types

The three broadest categories of symmetry are *total symmetry, partial symmetry,* and *strong symmetry*. A totally symmetric *n*-input Boolean function has symmetry group S_n. Intuitively, a partially symmetric function is a function that is totally symmetric on one or more subsets of inputs. The symmetry groups of partially and totally symmetric functions are generated by their 2-cycles making it possible to detect such symmetries using a constructive approach that detects symmetries between pairs of variables.

Strong symmetry is a catch-all term that includes any type of symmetry that is neither total nor partial. There are many distinctly different types of strong symmetry, and the number of different types grows quickly with respect to the number of inputs.

The function $x_0x_1x_2x_3$ is totally symmetric, the function $x_0x_1x_2+x_3$ is partially symmetric, and the function $x_0x_1+x_2x_3$ is strongly symmetric, because the variables x_0 and x_1 can be swapped as a set with the variables x_2 and x_3. Figure 13 Gives the symmetry groups for these three functions.

$$x_0x_1x_2x_3 \ - \ S_4$$
$$x_0x_1x_2 + x_3 \ - \ \{I,(0,1),(0,2),(1,2),(0,1,2),(0,2,1)\}$$
$$x_0x_1 + x_2x_3 \ -$$
$$\{I,(0,1),(2,3),(0,1)(2,3),(0,2)(1,3),(0,3)(1,2),(0,2,1,3),(0,3,1,2)\}$$

Figure 13. Symmetry groups for three functions.

3.3. Orbits

There have been some attempts to categorize permutation groups by tracking the locations where each input can appear. This is somewhat successful because in the function $x_0x_1x_2x_3$, the variable x_0 can appear in any of the four positions, 0, 1, 2, or 3, while in the function $x_0x_1x_2 + x_3$, x_0 can appear only in positions 0, 1 and 2. However this is only a partial success because in the function $x_0x_1 + x_2x_3$, the variable x_0 can appear in any position, even though the symmetry of this function is very different from that of $x_0x_1x_2x_2$.

Nevertheless, it is worthwhile to pursue this idea in a formal way. The concept of an *orbit* captures the ideas we have been describing. Remember that a permutation , p , is a function from the set $\{0,1,...,n\}$ to itself. Suppose H is a subgroup of S_n. Two elements $a,b \in \{0,1,...,n\}$ belong to the same orbit of H if there is a permutation $p \in H$ such that $p(a) = b$. It's easy to show that if R_1 and R_2 are two orbits of H then either $R_1 = R_2$ or $R_1 \cap R_2 = \phi$. Figure 14 gives the orbits of the symmetry groups of the functions shown in Figure 13.

$$x_0x_1x_2x_3 \ - \ \{0,1,2,3\}$$
$$x_0x_1x_2 + x_3 \ - \ \{0,1,2\},\{3\}$$
$$x_0x_1 + x_2x_3 \ - \ \{0,1,2,3\}$$

Figure 14. First-level orbits.

When a subgroup has only a single orbit, as with the symmetry groups of $x_0x_1x_2x_3$ and $x_0x_1 + x_2x_3$, the subgroup is said to be *transitive*. Orbits cannot be used to distinguish one transitive subgroup from another. When two subgroups are transitive, higher-level orbits can often be used to distinguish between them. The orbits shown in Figure 14 are only the first-level orbits. Second-level orbits are computed using ordered pairs of inputs, third-level orbits are computed using ordered triples, and so forth. When computing the second-level orbits, ordered pairs containing identical elements like (0,0) or (1,1) are omitted because these are partitioned identically with the first-level orbits. By the same token, when considering the third-level orbits, only those ordered triples that have three distinct elements are used. This simplifies the process of computing the higher-level orbits, but they are still more computationally intensive than first-level orbits. Figure 15 gives the second-level orbits of the three functions shown in Figure 13.

$$x_0x_1x_2x_3 \; - $$
$$\{(0,1),(1,0),(0,2),(2,0),(0,3),(3,0),(1,2),(2,1),(1,3),(3,1),(2,3),(3,2)\}$$

$$x_0x_1x_2 + x_3 \; - \; \begin{array}{l} \{(0,1),(1,0),(0,2),(2,0),(1,2),(2,1)\} \\ \{(0,3),(1,3),(2,3)\} \\ \{(3,0),(3,1),(3,2)\} \end{array}$$

$$x_0x_1 + x_2x_3 \; - \; \begin{array}{l} \{(0,1),(1,0),(2,3),(3,2)\} \\ \{(0,2),(1,2),(0,3),(1,3),(2,0),(2,1),(3,0),(3,1)\} \end{array}$$

Figure 15. Second-level orbits.

As Figure 15 shows, second-level orbits can distinguish between $x_0x_1x_2x_3$ and $x_0x_1 + x_2x_3$. When there is only one second-level orbit, the subgroup is said to be 2-transitive. Obviously, second-level orbits cannot distinguish between two 2-transitive subgroups. However, higher-level transitivities are much rarer than first and second-level transitivities so higher-level orbits are usually extremely effective in distinguishing one type of symmetry from another.

Although higher-level transitivities are an effective means for distinguishing one permutation group from another, they are not always an effective tool for analyzing the symmetries of Boolean functions. The reason for this is not every permutation group can be the symmetry group of a Boolean function. For example, S_3 has six subgroups $\{I\}$, $\{I,(0,1)\}$, $\{I,(0,2)\}$, $\{I,(1,2)\}$, $\{I,(0,1,2),(0,2,1)\}$, and S_3 itself. The function $x_0 + x_1 x_2'$ has symmetry group $\{I\}$, the functions $x_0 x_1 + x_2$, $x_0 x_2 + x_1$ and $x_0 + x_1 x_2$ have the symmetry groups $\{I,(0,1)\}$, $\{I,(0,2)\}$ and $\{I,(1,2)\}$ respectively, and $x_0 x_1 x_2$ has symmetry group S_3. But no 3-input Boolean function has the symmetry group $\{I,(0,1,2),(0,2,1)\}$. There is no way to determine this by analyzing the orbits of the subgroups.

To show that $\{I,(0,1,2),(0,2,1)\}$ cannot be the symmetry group of any Boolean function it is necessary to focus on the input values of the functions rather than on the input variables. A three-input Boolean function has eight possible inputs, 000, 001, 010, 011, 100, 101, 110, and 111. If we number the positions of these vectors 0, 1, and 2, we can apply permutations to them in the same manner as we have been doing with input variables. We can also compute orbits with respect to the Boolean inputs. We will consider two vectors to belong to the same orbit if there is a permutation that can transform one vector into another. These orbits are called *Boolean orbits* to distinguish them from ordinary orbits. Figure 16 gives the Boolean orbits for the subgroups of S_3.

$\{I\}$ – {000}, {001}, {010}, {011}, {100}, {101}, {110}, {111}.
$\{I,(0,1)\}$ – {000}, {001}, {010,100}, {011,101}, {110}, {111}.
$\{I,(0,2)\}$ – {000}, {001,100}, {010}, {011,110}, {101}, {111}.
$\{I,(1,2)\}$ – {000}, {001,010}, {011}, {100}, {101,110}, {111}.
$\{I,(0,1,2),(0,2,1)\}$ – {000}, {001,010,100}, {011,101,110}, {111}.
S_3 – {000}, {001,010,100}, {011,101,110}, {111}.

Figure 16. The Boolean orbits of S_3 subgroups.

As Figure 16 shows, the Boolean orbits of $\{I,(0,1,2),(0,2,1)\}$ and S_3 are identical. Now, suppose we are given a Boolean function f with symmetry group H. If two vectors v_1 and v_2 are in the same Boolean orbit, then there must be a permutation $p \in H$ that transforms v_1 into v_2. Because

f is invariant with respect to p, f must map both v_1 and v_2 to the same value. Therefore, any Boolean function that is invariant with respect to the subgroup $\{I,(0,1,2),(0,2,1)\}$ must be totally symmetric.

Because Boolean orbits partition the truth table of a function, they can be used to completely characterize its symmetry. An effective, but computationally intensive, method of determining the symmetry group of any n-input Boolean function, f, is to enumerate all subgroups of S_n and compute their Boolean orbits. Then partition the truth-table of f into ones and zeros and determine the largest group that leaves f invariant. This approach is practical only for small values of n.

A more efficient approach is to pre-compute the Boolean orbits of the groups that define the symmetries in which we are interested. Each Boolean orbit can be expressed as a characteristic function indicating which vectors belong to the orbit. These functions can then be stored in some compact form and used to determine whether or not a function has the desired symmetry.

The concept of function implication is used to test for a particular symmetry. The Boolean function f implies g if $g(x)$=1 whenever $f(x)$=1. If $\{C_1,C_2,...,C_k\}$ is the set of functions derived from the Boolean orbits of a group G, and if each of the C_i implies either f or its inverse $\overline{f}$, then G is a subgroup of the symmetry group of f. The number of functions, C_i, is generally much smaller than the number of input vectors or the size of the symmetry group. For example for 5-input functions, the group S_5 has 120 members, 32 input vectors, but only 6 Boolean orbits. Deriving the set of functions $\{C_1,C_2,...,C_k\}$, is computationally intensive, but the computation is off-line and need be done only once.

3.4. Forbidden Subgroups

If a subgroup H of S_n cannot be the permutation group of any n-input Boolean function, we call H a *forbidden subgroup*. Identifying the forbidden subgroups of S_n is a laborious process, but there are certain subgroups that can be eliminated immediately. Recall that permutations can be classified as even or odd permutations depending on whether they can be decomposed into an even or odd number of two-cycles. It is easy to see that the product of two even permutations or two odd permutations must be an even permutation, and the product of an even and an odd permutation must be an odd permutation. Let A_n be the set of all the even permutations of S_n. Because the product of any two even permutations must be another even permutation, A_n is closed under multiplication, and must be a subgroup of S_n. The subgroup A_n is called the *alternating group of degree n*. A_3 is the subgroup $\{I,(0,1,2),(0,2,1)\}$, because $I=(0,1)(0,1)$, $(0,1,2)=(0,1)(0,2)$, and $(0,2,1)=(0,2)(0,1)$, and the other three elements of S_3, $(0,1)$, $(0,2)$, and $(1,2)$ are obviously odd permutations. We already showed that A_3 cannot be the symmetry group of any function, because any function that is invariant with respect to A_3 will also be invariant with respect to S_3. This is also true for any n greater than 2. The intuitive reason for this is that the symmetry of a Boolean function depends on permutations of Boolean input vectors, and if n is greater than 2, then any vector of n elements must have at least two elements that are identical. Swapping these two elements will not change the vector, so we can add an extra 2-cycle that swaps these two elements to any permutation we apply to the vector.

We formalize this notion in the following lemma.

Lemma 1. Given two *n*-element vectors, v_1 and v_2, with $n>2$ and an odd permutation p_o such that $p_o(v_1)=v_2$, there is an even permutation p_e such that $p_e(v_1)=v_2$.

Proof. If $n>2$, v_1 must have at least two distinct elements that are equal. Suppose these elements are in positions i and j ($i\neq j$). Let $p_e=(i,j)p_o$. The permutation p_e is an even permutation, and because exchanging the elements in positions i and j will not change v_1, $p_e(v_1)=((i,j)p_o)(v_1)=p_o((i,j)(v_1))=p_o(v_1)=v_2$.

We can use this lemma to prove something more important, namely that if $n>2$ then A_n cannot be the symmetry group of any Boolean function.

Theorem 1. If $n > 2$, then A_n and S_n have the same Boolean orbits.

Proof. Suppose v_1 and v_2 are in the same Boolean orbit of A_n. Then there must be a permutation $p \in A_n$ such that $p(v_1) = v_2$. Since A_n is a subgroup of S_n, $p \in S_n$, so v_1 and v_2 are in the same Boolean orbit of S_n. Now suppose that v_1 and v_2 are in the same Boolean orbit of S_n. This means that there is a $p \in S_n$ such that $p(v_1) = v_2$. If p is even then p is already in A_n. Suppose that p is odd. If this is the case, then by Lemma 1, there is an even permutation p_e such that $p_e(v_1) = v_2$. Thus in either case there is a permutation $p' \in A_n$, equaling either p or p_e, such that $p'(v_1) = v_2$. Thus v_1 and v_2 are in the same Boolean orbit of A_n, and A_n and S_n have the same Boolean orbits.

Two permutation groups represent the same kind of symmetry if their Boolean orbits are identical. If two groups G_1 and G_2 have the same Boolean orbits then either one is a subgroup of the other, or there is a third group G_3 which has both G_1 and G_2 as subgroups and has the same Boolean orbits as G_1 and G_2. In fact, $G_3 = \langle G_1 \cup G_2 \rangle$, since G_3 is all of the permutations that can be created by multiplying the elements G_1 and G_2 together, nothing new is introduced in G_3. This can be formalized as Theorem 2.

Theorem 2. Suppose two subgroups G_1 and G_2 of S_n have the same Boolean orbits. Then $G_3 = \langle G_1 \cup G_2 \rangle$ has the same Boolean orbits as G_1 and G_2.

Proof. Let a be an element of G_3, and let v_1 be any *n*-element Boolean vector. If $a(v_1) = v_2$, we must show that v_1 and v_2 are in the same Boolean orbit of G_1 (and G_2). The element a must be of the form $a_1 a_2 ... a_k$. We will proceed by induction on k. If $k = 1$, then $a_k \in G_1$ or $a_k \in G_2$, and since the Boolean orbits of G_1 and G_2 are identical, v_1 and v_2 are in the same

Boolean orbit of G_1. Now suppose $a = a_1 a_2 ... a_k$, $k > 1$. $a(v_1) = (a_2 ... a_k)(a_1(v_1))$. Let $v_3 = a_1(v_1)$. Because either $a_1 \in G_1$ or $a_1 \in G_2$, v_3 and v_1 must be in the same Boolean orbit of G_1 (and G_2). But $v_2 = (a_2 ... a_k)(v_3)$ The vectors v_3 and v_2 must be in the same Boolean orbit of G_1 by the inductive hypothesis, and because belonging to the same Boolean orbit is an equivalence relation, v_1 and v_2 must be in the same Boolean orbit of G_1. Thus if v_1 and v_2 are in the same Boolean orbit of G_3, they must be in the same Boolean orbit of G_1. Because G_1 and G_2 are subgroups of G_3, The orbits of G_3 cannot be smaller than the Boolean orbits of G_1 (or G_2). If v_1 and v_2 are in the same Boolean orbit of G_1 then there is a $p \in G_1$ such that $p(v_1) = v_2$. But p is also an element of G_3, so v_1 and v_2 are in the same Boolean orbit of G_3.

Theorem 2 also includes the case where one subgroup is a subset of the other since if $G_1 \subseteq G_2$ then $\langle G_1 \cup G_2 \rangle = G_2$.

The point of Theorem 2 is to show that if there are several subgroups of S_n that have the same Boolean orbits, then there must be a largest one that contains all the others. Only the largest of these can be the symmetry group of a Boolean function, because the symmetry group of a Boolean function must include *all* the permutations that leave the function invariant.

$$\{0000\}$$
$$\{0001,0010,0100,1000\}$$
$$\{0110,1001\}$$
$$\{0101,1010\}$$
$$\{0011,1100\}$$
$$\{0111,1011,1101,1110\}$$
$$\{1111\}$$

Figure 17. The Boolean orbits of the Klein 4-group.

Even when a symmetry group has Boolean orbits that are different from every other permutation group of the same degree, it can still fail to be the permutation group of any Boolean function. Consider the subgroup of S_4, $\{I, (0,1)(2,3), (0,2)(1,3), (0,3)(1,2)\}$. This group is known as the *Klein 4-group*. The Klein 4-group cannot be the symmetry group of any 4-input Boolean function, but the reason is neither simple nor obvious.

The problem lies with the Boolean orbits which are shown in Figure 17.

Consider the three Boolean orbits, $\{0110,1001\}$, $\{0101,1010\}$, and $\{0011,1100\}$. Figure 18 shows three subgroups of S_4, G_1, G_2 and G_3. The Klein 4-group is a subgroup of each of these three groups.

$$G_1 = \{\mathrm{I},(2,3),(0,1),(0,1)(2,3),(0,2)(1,3),(0,2,1,3),(0,3,1,2),(0,3)(1,2)\}$$

$$\{0000\}$$
$$\{0001,0010,0100,1000\}$$
$$\{0110,1001,0101,1010\}$$
$$\{0011,1100\}$$
$$\{0111,1011,1101,1110\}$$
$$\{1111\}$$

$$G_2 = \{\mathrm{I},(1,2),(0,1)(2,3),(0,1,3,2),(0,2,3,1),(0,2)(1,3),(0,3),(0,3)(1,2)\}$$

$$\{0000\}$$
$$\{0001,0010,0100,1000\}$$
$$\{0110,1001\}$$
$$\{0101,1010,0011,1100\}$$
$$\{0111,1011,1101,1110\}$$
$$\{1111\}$$

$$G_3 = \{\mathrm{I},(1,3),(0,1)(2,3),(0,1,2,3),(0,2),(0,2)(1,3),(0,3,2,1),(0,3)(1,2)\}$$

$$\{0000\}$$
$$\{0001,0010,0100,1000\}$$
$$\{0101,1010\}$$
$$\{0110,1001,0011,1100\}$$
$$\{0111,1011,1101,1110\}$$
$$\{1111\}$$

Figure 18. Three groups with their Boolean orbits.

The Boolean orbits of groups G_1, G_2 and G_3 differ only slightly from those of the Klein 4-group. The orbits are identical, except for combining two of the three orbits $\{0110,1001\}$, $\{0101,1010\}$, and $\{0011,1100\}$. The group G_1 combines $\{0110,1001\}$ and $\{0101,1010\}$. The group G_2 combines $\{0011,1100\}$ and $\{0101,1010\}$. The group G_3 combines $\{0110,1001\}$ and $\{0011,1100\}$. If the Klein 4-group leaves a Boolean function f invariant, then at least two of the three orbits $\{0110,1001\}$, $\{0101,1010\}$, and $\{0011,1100\}$ must be assigned the same value by f. This means that if the Klein 4-group leaves f invariant then at least one of G_1, G_2 or G_3 must also leave f invariant, and the Klein 4-group cannot be the symmetry group of f.

Because Boolean functions can have the only two different values, there aren't enough function values to distinguish the three critical Boolean orbits. If we were to consider functions with 2-bit outputs, then it would be possible for the Klein 4-group to be the symmetry group of such a function.

3.5. Some Recognized Types of Symmetry

As stated above, all symmetries can be categorized as total symmetry, partial symmetry or strong symmetry. The class of strong symmetries has been further subdivided into several different recognizable types of symmetry. This subdivision is by no means exhaustive, so there are many types of symmetries that are simply classified as "other."

One of the simplest types of symmetry is *hierarchical symmetry*. Examples of functions with hierarchical symmetry are $f = x_0x_1 + x_2x_3$ and $g = x_0x_1x_2 + x_3x_4x_5 + x_6x_7x_8$. In both of these functions we can exchange certain selected pairs of variables. For function f we can exchange x_0 and x_1 or x_2 and x_3. In addition, we can exchange the pair (x_0, x_1) with the pair (x_2, x_3). Hierarchical symmetry is characterized by a set of partial symmetries combined with the ability to exchange one set of partially symmetric variables with another such set. Hierarchical symmetries can be nested to an arbitrary depth.

Technically, a hierarchical symmetry can be characterized as a special type of group product known as a *wreath product*. Suppose we have a function f with k sets of input variables all of the same size. When viewed in isolation, each one of these sets has the same permutation group, G. In addition, it is possible to permute the sets among one another, using permutation group H. The permutation group of f is called the *wreath product* of G and H. In hierarchical symmetry, G and H must be isomorphic to S_i and S_j for some i and j. It is not necessary that i be equal to j. The symmetry group of the function $f = x_0x_1 + x_2x_3$ is the wreath product of S_2 with itself, and the symmetry group of $g = x_0x_1x_2 + x_3x_4x_5 + x_6x_7x_8$ is the wreath product of S_3 with itself. Because hierarchical symmetry, as it is currently understood, restricts the factors to isomorphic copies of S_i, it barely scratches the surface of wreath products.

For some functions, it is possible to permute sets of variables that are *not* partially symmetric. An example of a function with this type of symmetry is $h = x_0'x_1 + x_2'x_3$. It is possible to exchange the pair of variables (x_0, x_1) with the ordered pair (x_2, x_3), but it is not possible to exchange the elements of either ordered pair. This type of symmetry is sometimes known as *set symmetry*. In set symmetry, the inputs of a function can be partitioned into two or more sets of variables. Any permutation on one set of variables forces a permutation on the other sets. This is another way of saying that the permutation group of the function consists of permutations that are disjoint cycles. For example, the permutation group of $h = x_0'x_1 + x_2'x_3$ is $\{I, (0,2)(1,3)\}$.

In set symmetry, the sets are usually of the same size. The reason for this can be understood by imagining that we have a 5-input function, f, with input variables x_0 through x_4. Suppose that it is possible to exchange variables x_0 and x_1, but if I do so I am also required to rotate the variables x_2, x_3, and x_4. This operation is characterized by the permutation $(0,1)(2,3,4)$. If I do this operation twice the result is the permutation $(2,4,3)$ if I do it three times, the result is $(0,1)$, four times and the result is $(2,3,4)$, five times and the result is $(0,1)(2,4,3)$, six times and the result is I. All of these permutations are in the permutation group of f which is $\{I,(0,1)(2,3,4),(2,4,3),(0,1),(2,3,4),(0,1)(2,4,3)\}$. This is the same as the permutation group of the function $g = x_0x_1 + x_2x_3x_4$, which is partially symmetric. Because the two sets are of different sizes, the components of the permutation disappear selectively when the permutation is performed several times. This allows the remaining components to be used in isolation, negating the effect of set symmetry.

Two related types of symmetry are rotational and dihedral symmetry. Both of these are related to the symmetry of regular polygons. Imagine a coin in the shape of a regular polygon that is to be inserted into a hole of the same shape, as shown in Figure 19.

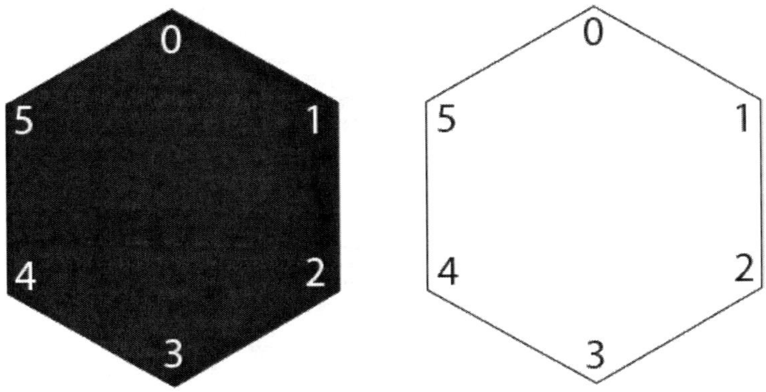

Figure 19. A Hexagonal Peg in a Hexagonal Hole.

The coin can be inserted so the numbers match up, or it can be rotated so zero matches up with one of the five other numbers. These rotations correspond to the set permutations $\{I,(0,1,2,3,4,5),(0,2,4)(1,3,5),(0,3)(1,4)(2,5),(0,4,2)(1,5,3),(0,5,4,3,2,1)\}$. These permutations are simply the single rotation $(0,1,2,3,4,5)$ performed one, two, three, four, five and six times. If we insist that the coin be placed heads-up, then this set of permutations gives all permissible positions for the coin. This group is called R_6 or the *rotational group* on six elements. The rotational group on n elements is generated in a similar fashion.

If it is possible to turn the coin over, then we double the size of the group. To flip the coin over, we add the permutation (1,5)(2,4). Combining this permutation with the rotations will give us a total of twelve permutations. The remaining permutations are (0,2)(3,5), (1,3)(0,4), (0,1)(2,5)(3,4), (0,3)(1,2)(4,5), and (0,5)(1,4)(2,3). This group is called D_{12} or the *dihedral group* of order 12. The general dihedral group is designated as D_{2n} to signify that it is a group of size $2n$ generated by a polygon with n sides. The symmetry group of $f = x_0x_1 + x_2x_3$ is D_8. The function f has the same symmetry as a square coin. The two variables x_0 and x_1 represent two corners that are diagonally across from one another.

Rotational symmetry cannot exist for any function with fewer than six inputs. This is because any sequence of five or fewer digits can be reversed in sequence by rotating it a sufficient number of times. This means that if $n < 6$ then R_n and D_{2n} will have the same Boolean orbits. Thus any function of five or fewer inputs that is rotationally symmetric will also be dihedrally symmetric. For functions of six or more inputs, the bit sequence 110010 cannot be reversed by rotating it. Thus rotational symmetry may exist for functions of six or more inputs.

3.6. Enumerating Permutation Groups

The enumeration problem is trivial for S_2 and S_3. S_2 consists of the identity and the permutation (0,1), giving us the subgroups S_2 and $\{I\}$ so a 2-input function is either non-symmetric or totally symmetric. For S_3 the situation is only slightly more complex. The subgroups are $\{I\}$, $\{I,(0,1)\}$ and its two conjugates, A_3 and S_3. A_3 is forbidden so any three input function is either totally symmetric, partially symmetric in two variables or non-symmetric.

For S_4, things begin to get more interesting. The various subgroups are listed in Figure 20. To save space we will list only one member of a set of conjugate groups. As expected, we have total symmetry, and three types of partial symmetry, one in two variables, one in three variables, and one in two pairs of two variables. There is also dihedral symmetry, which we have encountered previously, and a set symmetry which allows us to exchange two particular inputs as long as we exchange the other two at the same time.

	Group	Symmetry
1	I	Non-symmetric
2	I, (0,1)	Partial symmetry in two variables: $x_0x_1 + x_2 + x_3'$
3	I, (0,1)(2,3)	Exchange two pairs of variables, but not the individual variables. $x_0x_1' + x_2x_3'$
4	A_3 (conjugate to)	Forbidden
5	I, (0,1), (2,3), (0,1)(2,3)	Double partial symmetry in two pairs of variables: $x_0x_1 + x_2 + x_3$
6	I, (0,1)(2,3), (0,2)(1,3), (0,3)(1,2)	Forbidden for single-output functions.
7	R_4	Forbidden
8	S_3 (conjugate to)	Partial symmetry in three variables: $x_0x_1x_2 + x_3$
9	D_8	Hierarchical symmetry $x_0x_1 + x_2x_3$
10	A_4	Forbidden
11	S_4	Total symmetry

Figure 20. The subgroups of S_4 .

For five inputs, we have a repetition of the types of symmetry encountered with four inputs, plus a couple of new additions. With five inputs, the Klein 4-group is no longer forbidden, because we can use the fifth bit to distinguish the appropriate Boolean orbits. Also we now have two dihedral symmetries, a "partial" dihedral symmetry in four variables and a total dihedral symmetry in five variables. There is also a partial symmetry in four variables.

Of these new symmetries, the Klein 4-group symmetry is the most interesting. Functions exhibiting this symmetry are complex, in the sense that they have many prime implicants, and functions that possess this type of symmetry are extremely difficult to identify.

Based on the information obtained from the subgroups of S_4 and S_5 we might be tempted to conclude that our existing categories of symmetry encompass virtually all types of symmetry. This conclusion would be wrong, as Figures 21 through 27 show. Functions of six variables exhibit many new types of symmetry that have yet to be classified, and some that will probably never fit neatly into any easily defined category. In point of fact, our experience with the Klein 4-group in S_5 should have warned us that exotic, difficult to classify, types of symmetry exist. Again, any type of symmetry that exists for 2, 3, 4, and 5-input functions also exists for 6-input functions.

	Group	Symmetry
1	I	Non symmetric
2	I, (0,1)	Partial symmetry in 2 variables
3	I, (0,1)(2,3)	Exchange two pairs of variables, but not the individual variables.
4	A_3	Forbidden
5	I, (2,3), (0,1), (0,1)(2,3)	Double partial symmetry
6	I, (0,1)(2,3), (0,2)(1,3), (0,3)(1,2)	Klein 4-group. No longer forbidden.
7	R_4	Forbidden
8	R_5	Forbidden
9	I, (3,4), (0,1,2), (0,1,2)(3,4), (0,2,1), (0,2,1)(3,4)	Forbidden
10	S_3	Partial symmetry in three variables.
11	I, (1,2)(3,4), (0,1)(3,4), (0,1,2), (0,2,1), (0,2)(3,4)	Forbidden
12	D_8	Dihedral symmetry in four variables
13	D_{10}	Dihedral symmetry
14	I, (3,4), (1,2), (1,2)(3,4), (0,1), (0,1)(3,4), (0,1,2), (0,1,2)(3,4), (0,2,1), (0,2,1)(3,4), (0,2), (0,2)(3,4)	Partial symmetry in one set of 3 variables and one set of 2 variables.
15	A_4	Forbidden
16	I, (1,2,4,3), (1,3,4,2), (1,4)(2,3), (0,1)(2,4), (0,1,2,3,4), (0,1,3,2), (0,1,4,3), (0,2,3,1), (0,2)(3,4), (0,2,4,1,3), (0,2,1,4), (0,3,4,1), (0,3,2,4), (0,3)(1,2), (0,3,1,4,2), (0,4,3,2,1), (0,4,2,3), (0,4,1,2), (0,4)(1,3)	Forbidden
17	S_4	Partial symmetry in 4 variables
18	A_5	Forbidden
19	S_5	Total symmetry.

Figure 21. The subgroups of S_5 .

Subgroup 4 of Figure 21 gives us a new set symmetry, in which it is possible to exchange three variables for three different variables, but does not permit the set of three variables to be rearranged.

Subgroup 6 allows a set of three variables to be rotated sequentially, as long as the other set of three variables is rotated in precisely the same way. Any rearrangement of the variables must affect all six input variables. The symmetry is identical to that of a rotating cylinder that has been divided into thirds at each end. It is not permissible to flip the cylinder from end to end, but it is permissible to rotate it 120° in either direction.

Group 7 gives us a partial symmetry in two variables combined with a set symmetry identical to that of Group 2 in the other four variables.

Group 9 is the first of the unclassifiable symmetries. A function with this type of symmetry will have three pairs of variables any two of which can be exchanged simultaneously as long as the third pair is left intact. The inputs in a particular pair *may not* be exchanged with one another. At first glance, Group 9 appears to be conjugate to the Klein 4-group. However, this is not the case. In Group 9, every 2-cycle appears twice, while in the Klein 4-group no 2-cycle appears more than once.

Group 10 is also unclassifiable. It is a combination of a partial symmetry in two variables with the Klein 4-group. However the two groups are combined in such a way so that neither group is allowed to act independently. Two of the Klein-4 group operations require a simultaneous exchange of variables 0 and 1. The third Klein 4-group operation must be performed without exchanging variables 0 and 1.

Group 13 links 4-variable rotational symmetry with a partial symmetry in the other two variables. An odd permutation of the four variables forces an exchange of the other two.

Group 19 is related to Group 6 in that it represents the symmetry of a rotating cylinder divided into thirds. In addition to the rotation, it is possible to swap the cylinder end for end, as long as one end is twisted into reverse order. (Of course, it is impossible to do this with a physical object.)

Group 20 represents rotational symmetry which is no longer forbidden.

Group 25 represents a dihedral symmetry in the first four variables, linked with a partial symmetry in the last two variables. The groups are not permitted to operate independently. The two group generators (0,1,2,3) and (0,1)(2,3) both force a swap of variables 4 and 5. The remaining group elements are the closure of the generators (0,1,2,3)(4,5) and (0,1)(2,3)(4,5).

	Group	Symmetry
1	I	Nonsymmetric
2	I,(1,2)	Partial symmetry in 2 variables
3	I, (0,1)(2,3)	Exchange two sets of variables, but not the variables themselves
4	I, (0,1)(2,3)(4,5)	Exchange three sets of two variables, but not the variables themselves
5	A_3	Forbidden
6	I, (0,1,2)(3,4,5), (0,2,1)(3,5,4)	Rigid rotating triangular block
7	I, (4,5), (0,1)(2,3), (0,1)(2,3)(4,5)	Set symmetry combined with partial symmetry
8	I, (2,3), (0,1), (0,1)(2,3)	Double partial symmetry
9	I, (2,3)(4,5), (0,1)(4,5), (0,1)(2,3)	Three pairs of inputs, any two of which can be exchanged.
10	I, (2,4)(3,5), (0,1)(2,3)(4,5), (0,1)(2,5)(3,4)	Two Klein 4-group operations force an exchange of vars 0 and 1. The third does not permit the exchange.
11	I, (0,1)(2,3), (0,2)(1,3), (0,3)(1,2)	Klein-4 symmetry
12	R_4	Forbidden
13	I, (0,1,2,3)(4,5), (0,2)(1,3), (0,3,2,1)(4,5)	4-variable Rotational with exchange of other two variables
14	R_5	Forbidden
15	I, (4,5), (0,1,2), (0,1,2)(4,5), (0,2,1), (0,2,1)(4,5)	Forbidden
16	S_3	Partial-3
17	I, (1,2)(4,5), (0,1)(3,4), (0,1,2)(3,4,5), (0,2,1)(3,5,4), (0,2)(3,5)	Two Interlocked S_3 's
18	I, (1,2)(3,4), (0,1)(3,4), (0,1,2), (0,2,1), (0,2)(3,4)	Forbidden
19	I, (0,1,2)(3,4,5), (0,2,1)(3,5,4), (0,3)(1,5)(2,4), (0,4)(1,3)(2,5), (0,5)(1,4)(2,3)	Rotating triangular block. End swap OK with one-end reversal.
20	R_6	Rotational symmetry
21	I, (4,5), (2,3), (2,3)(4,5), (0,1), (0,1)(4,5), (0,1)(2,3), (0,1)(2,3)(4,5)	Triple partial symmetry in two variables.
22	I, (4,5), (0,1)(2,3), (0,1)(2,3)(4,5), (0,2)(1,3), (0,2)(1,3)(4,5), (0,3)(1,2), (0,3)(1,2)(4,5)	Klein-4 plus partial-2
23	I, (4,5), (0,1,2,3), (0,1,2,3)(4,5), (0,2)(1,3), (0,2)(1,3)(4,5), (0,3,2,1), (0,3,2,1)(4,5)	Forbidden
24	D_8	Dihedral in 4
25	I, (1,3), (0,1)(2,3)(4,5), (0,1,2,3)(4,5), (0,2), (0,2)(1,3), (0,3,2,1)(4,5), (0,3)(1,2)(4,5)	Dihedral in 4 linked with partial in 2

Figure 22. The subgroups of S_6 Part 1.

26	I, (1,3)(4,5), (0,1)(2,3), (0,1,2,3)(4,5), (0,2)(4,5), (0,2)(1,3), (0,3,2,1)(4,5), (0,3)(1,2)	Dihedral in 4 linked with partial in two variables
27	I, (1,3)(4,5), (0,1)(2,3)(4,5), (0,1,2,3), (0,2)(4,5), (0,2)(1,3), (0,3,2,1), (0,3)(1,2)(4,5)	Forbidden
28	I, (3,4,5), (3,5,4), (0,1,2), (0,1,2)(3,4,5), (0,1,2)(3,5,4), (0,2,1), (0,2,1)(3,4,5), (0,2,1)(3,5,4)	Forbidden
29	D_{10}	Dihedral in 5
30	I, (3,4), (1,2), (1,2)(3,4), (0,1), (0,1)(3,4), (0,1,2), (0,1,2)(3,4), (0,2,1), (0,2,1)(3,4), (0,2), (0,2)(3,4)	3-2 partial symmetry
31	A_4	Forbidden
32	I, (1,5)(2,3), (0,1,2)(3,4,5), (0,1,3)(2,4,5), (0,2,1)(3,5,4), (0,2,5)(1,4,3), (0,3,1)(2,5,4), (0,3,5)(1,4,2), (0,4)(2,3), (0,4)(1,5), (0,5,3)(1,2,4), (0,5,2)(1,3,4)	Pair swap with rotation in the pairs.
33	D_{12}	Dihedral symmetry
34	I, (4,5), (1,3), (1,3)(4,5), (0,1)(2,3), (0,1)(2,3)(4,5), (0,1,2,3), (0,1,2,3)(4,5), (0,2), (0,2)(4,5), (0,2)(1,3), (0,2)(1,3)(4,5), (0,3,2,1), (0,3,2,1)(4,5), (0,3)(1,2), (0,3)(1,2)(4,5)	Dihedral in 4 with independent partial in 2
35	I, (4,5), (3,4), (3,4,5), (3,5,4), (3,5), (0,1,2), (0,1,2)(4,5), (0,1,2)(3,4), (0,1,2)(3,4,5), (0,1,2)(3,5,4), (0,1,2)(3,5), (0,2,1), (0,2,1)(4,5), (0,2,1)(3,4), (0,2,1)(3,4,5), (0,2,1)(3,5,4), (0,2,1)(3,5)	Forbidden
36	I, (3,4,5), (3,5,4), (1,2)(4,5), (1,2)(3,4), (1,2)(3,5), (0,1)(4,5), (0,1)(3,4), (0,1)(3,5), (0,1,2), (0,1,2)(3,4,5), (0,1,2)(3,5,4), (0,2,1), (0,2,1)(3,4,5), (0,2,1)(3,5,4), (0,2)(4,5), (0,2)(3,4), (0,2)(3,5)	Forbidden
37	I, (1,3,5), (1,5,3), (0,1)(2,3)(4,5), (0,1,2,3,4,5), (0,1,4,5,2,3), (0,2,4), (0,2,4)(1,3,5), (0,2,4)(1,5,3), (0,3,2,5,4,1), (0,3,4,1,2,5), (0,3)(1,4)(2,5), (0,4,2), (0,4,2)(1,3,5), (0,4,2)(1,5,3), (0,5,4,3,2,1), (0,5)(1,2)(3,4), (0,5,2,1,4,3)	Forbidden
38	I, (1,2,4,3), (1,3,4,2), (1,4)(2,3), (0,1)(2,4), (0,1,2,3,4), (0,1,3,2), (0,1,4,3), (0,2,3,1), (0,2)(3,4), (0,2,4,1,3), (0,2,1,4), (0,3,4,1), (0,3,2,4), (0,3)(1,2), (0,3,1,4,2), (0,4,3,2,1), (0,4,2,3), (0,4,1,2), (0,4)(1,3)	Forbidden
39	I, (3,4), (1,2,5), (1,2,5)(3,4), (1,5,2), (1,5,2)(3,4), (0,1)(2,5), (0,1)(2,5)(3,4), (0,1,2), (0,1,2)(3,4), (0,1,5), (0,1,5)(3,4), (0,2,1), (0,2,1)(3,4), (0,2,5), (0,2,5)(3,4), (0,2)(1,5), (0,2)(1,5)(3,4), (0,5,1), (0,5,1)(3,4), (0,5,2), (0,5,2)(3,4), (0,5)(1,2), (0,5)(1,2)(3,4)	Forbidden
40	S_4	Partial symmetry in four variables

Figure 23. The subgroups of S_6 Part 2.

41	I, (2,3)(4,5), (1,2)(4,5), (1,2,3), (1,3,2), (1,3)(4,5), (0,1)(4,5), (0,1)(2,3), (0,1,2), (0,1,2,3)(4,5), (0,1,3,2)(4,5), (0,1,3), (0,2,1), (0,2,3,1)(4,5), (0,2)(4,5), (0,2,3), (0,2)(1,3), (0,2,1,3)(4,5), (0,3,2,1)(4,5), (0,3,1), (0,3,2), (0,3)(4,5), (0,3,1,2)(4,5), (0,3)(1,2)	S_4 in first 4 variables. Odd permutations require simultaneous swap of last two variables
42	I, (2,5), (1,4), (1,4)(2,5), (0,1,2)(3,4,5), (0,1,2,3,4,5), (0,1,5,3,4,2), (0,1,5)(2,3,4), (0,2,1)(3,5,4), (0,2,4,3,5,1), (0,2,1,3,5,4), (0,2,4)(1,3,5), (0,3), (0,3)(2,5), (0,3)(1,4), (0,3)(1,4)(2,5), (0,4,5,3,1,2), (0,4,5)(1,2,3), (0,4,2)(1,5,3), (0,4,2,3,1,5), (0,5,4,3,2,1), (0,5,1)(2,4,3), (0,5,4)(1,3,2), (0,5,1,3,2,4)	Keep the pairs (0,3), (1,4) and (2,5) diametrically opposed. No reflections.
43	I, (1,2)(3,5), (1,3)(2,5), (1,5)(2,3), (0,1)(4,5), (0,1,2)(3,4,5), (0,1,3)(2,4,5), (0,1,4,5)(2,3), (0,2,1)(3,5,4), (0,2)(3,4), (0,2,5)(1,4,3), (0,2,4,3)(1,5), (0,3,1)(2,5,4), (0,3)(2,4), (0,3,5)(1,4,2), (0,3,4,2)(1,5), (0,4)(2,3), (0,4)(1,2,5,3), (0,4)(1,3,5,2), (0,4)(1,5), (0,5,4,1)(2,3), (0,5,3)(1,2,4), (0,5,2)(1,3,4), (0,5)(1,4)	Complex interaction of several symmetries
44	I, (1,2,3,4), (1,3)(2,4), (1,4,3,2), (0,1)(2,4)(3,5), (0,1,2)(3,4,5), (0,1,4)(2,5,3), (0,1,5,3), (0,2,1)(3,5,4), (0,2,5,4), (0,2)(1,3)(4,5), (0,2,3)(1,5,4), (0,3,5,1), (0,3,4)(1,2,5), (0,3,2)(1,4,5), (0,3)(1,5)(2,4), (0,4,1)(2,3,5), (0,4,5,2), (0,4)(1,3)(2,5), (0,4,3)(1,5,2), (0,5)(2,4), (0,5)(1,2)(3,4), (0,5)(1,3), (0,5)(1,4)(2,3)	Forbidden
45	I, (4,5), (3,4), (3,4,5), (3,5,4), (3,5), (1,2), (1,2)(4,5), (1,2)(3,4), (1,2)(3,4,5), (1,2)(3,5,4), (1,2)(3,5), (0,1), (0,1)(4,5), (0,1)(3,4), (0,1)(3,4,5), (0,1)(3,5,4), (0,1)(3,5), (0,1,2), (0,1,2)(4,5), (0,1,2)(3,4), (0,1,2)(3,4,5), (0,1,2)(3,5,4), (0,1,2)(3,5), (0,2,1), (0,2,1)(4,5), (0,2,1)(3,4), (0,2,1)(3,4,5), (0,2,1)(3,5,4), (0,2,1)(3,5), (0,2), (0,2)(4,5), (0,2)(3,4), (0,2)(3,4,5), (0,2)(3,5,4), (0,2)(3,5)	Two independent 3-Partials
46	I, (2,4)(3,5), (1,3,5), (1,3)(2,4), (1,5,3), (1,5)(2,4), (0,1)(2,3)(4,5), (0,1)(2,5)(3,4), (0,1,2,3,4,5), (0,1,2,5,4,3), (0,1,4,5,2,3), (0,1,4,3,2,5), (0,2)(3,5), (0,2,4), (0,2)(1,3), (0,2,4)(1,3,5), (0,2)(1,5), (0,2,4)(1,5,3), (0,3,4,5,2,1), (0,3,2,5,4,1), (0,3)(1,2)(4,5), (0,3,4,1,2,5), (0,3,2,1,4,5), (0,3)(1,4)(2,5), (0,4,2), (0,4)(3,5), (0,4,2)(1,3,5), (0,4)(1,3), (0,4,2)(1,5,3), (0,4)(1,5), (0,5,4,3,2,1), (0,5,2,3,4,1), (0,5)(1,2)(3,4), (0,5,4,1,2,3), (0,5,2,1,4,3), (0,5)(1,4)(2,3)	Forbidden
47	I, (2,5)(3,4), (1,3,4), (1,3)(2,5), (1,4,3), (1,4)(2,5), (0,1)(2,3,5,4), (0,1)(2,4,5,3), (0,1,2,3)(4,5), (0,1,2,4)(3,5), (0,1,5,4)(2,3), (0,1,5,3)(2,4), (0,2)(3,4), (0,2,5), (0,2)(1,3), (0,2,5)(1,3,4), (0,2)(1,4), (0,2,5)(1,4,3), (0,3,2,1)(4,5), (0,3,5,1)(2,4), (0,3,5,4)(1,2), (0,3)(1,2,4,5), (0,3)(1,5,4,2), (0,3,2,4)(1,5), (0,4,2,1)(3,5), (0,4,5,1)(2,3), (0,4,5,3)(1,2), (0,4)(1,2,3,5), (0,4)(1,5,3,2), (0,4,2,3)(1,5), (0,5,2), (0,5)(3,4), (0,5,2)(1,3,4), (0,5)(1,3), (0,5,2)(1,4,3), (0,5)(1,4)	Forbidden
48	I, (4,5), (2,3), (2,3)(4,5), (1,2), (1,2)(4,5), (1,2,3), (1,2,3)(4,5), (1,3,2), (1,3,2)(4,5), (1,3), (1,3)(4,5), (0,1), (0,1)(4,5), (0,1)(2,3), (0,1)(2,3)(4,5), (0,1,2), (0,1,2)(4,5), (0,1,2,3), (0,1,2,3)(4,5), (0,1,3,2), (0,1,3,2)(4,5), (0,1,3), (0,1,3)(4,5), (0,2,1), (0,2,1)(4,5), (0,2,3,1), (0,2,3,1)(4,5), (0,2), (0,2)(4,5), (0,2,3), (0,2,3)(4,5), (0,2)(1,3), (0,2)(1,3)(4,5), (0,2,1,3), (0,2,1,3)(4,5), (0,3,2,1), (0,3,2,1)(4,5), (0,3,1), (0,3,1)(4,5), (0,3,2), (0,3,2)(4,5), (0,3), (0,3)(4,5), (0,3,1,2), (0,3,1,2)(4,5), (0,3)(1,2), (0,3)(1,2)(4,5)	Independent 4-2 partial

Figure 24. The subgroups of S_6 Part 3.

49	I, (2,5), (1,2)(4,5), (1,2,4,5), (1,4), (1,4)(2,5), (1,5,4,2), (1,5)(2,4), (0,1)(3,4), (0,1)(2,5)(3,4), (0,1,2)(3,4,5), (0,1,2,3,4,5), (0,1,3,4), (0,1,3,4)(2,5), (0,1,5,3,4,2), (0,1,5)(2,3,4), (0,2,1)(3,5,4), (0,2,4,3,5,1), (0,2)(3,5), (0,2,3,5), (0,2,1,3,5,4), (0,2,4)(1,3,5), (0,2)(1,4)(3,5), (0,2,3,5)(1,4), (0,3), (0,3)(2,5), (0,3)(1,2)(4,5), (0,3)(1,2,4,5), (0,3)(1,4), (0,3)(1,4)(2,5), (0,3)(1,5,4,2), (0,3)(1,5)(2,4), (0,4,3,1), (0,4,3,1)(2,5), (0,4,5,3,1,2), (0,4,5)(1,2,3), (0,4)(1,3), (0,4)(1,3)(2,5), (0,4,2)(1,5,3), (0,4,2,3,1,5), (0,5,4,3,2,1), (0,5,1)(2,4,3), (0,5,3,2), (0,5)(2,3), (0,5,4)(1,3,2), (0,5,1,3,2,4), (0,5,3,2)(1,4), (0,5)(1,4)(2,3)	Similar to 42, but with reflections
50	A_5	Forbidden
51	I, (2,4)(3,5), (1,2)(4,5), (1,2,5,3,4), (1,3,2,4,5), (1,3)(2,5), (1,4,3,5,2), (1,4)(2,3), (1,5)(3,4), (1,5,4,2,3), (0,1)(3,5), (0,1)(2,4), (0,1,2)(3,5,4), (0,1,2,3,4), (0,1,3,4,5), (0,1,3)(2,5,4), (0,1,4,5,2), (0,1,4)(2,5,3), (0,1,5,2,3), (0,1,5)(2,4,3), (0,2,1)(3,4,5), (0,2,5,4,1), (0,2)(3,4), (0,2,3,5,4), (0,2,1,3,5), (0,2,4,1,3), (0,2,3)(1,4,5), (0,2,5)(1,4,3), (0,2)(1,5), (0,2,4)(1,5,3), (0,3,1)(2,4,5), (0,3,2,5,1), (0,3)(4,5), (0,3,4,2,5), (0,3)(1,2), (0,3,5)(1,2,4), (0,3,1,4,2), (0,3,5,1,4), (0,3,2)(1,5,4), (0,3,4)(1,5,2), (0,4,3,2,1), (0,4,1)(2,3,5), (0,4,5,3,2), (0,4)(2,5), (0,4,5)(1,2,3), (0,4,3)(1,2,5), (0,4,2)(1,3,5), (0,4)(1,3), (0,4,2,1,5), (0,4,1,5,3), (0,5,4,3,1), (0,5,1)(2,3,4), (0,5)(2,3), (0,5,2,4,3), (0,5,3,1,2), (0,5,1,2,4), (0,5,2)(1,3,4), (0,5,4)(1,3,2), (0,5,3)(1,4,2), (0,5)(1,4)	All even permutations, but not forbidden. Too complex to describe.
52	I, (3,4), (2,5), (2,5)(3,4), (1,3), (1,3,4), (1,3)(2,5), (1,3,4)(2,5), (1,4,3), (1,4), (1,4,3)(2,5), (1,4)(2,5), (0,1)(2,3)(4,5), (0,1)(2,3,5,4), (0,1)(2,4,5,3), (0,1)(2,4)(3,5), (0,1,2,3)(4,5), (0,1,2,3,5,4), (0,1,2,4,5,3), (0,1,2,4)(3,5), (0,1,5,4,2,3), (0,1,5,4)(2,3), (0,1,5,3)(2,4), (0,1,5,3,2,4), (0,2), (0,2)(3,4), (0,2,5), (0,2,5)(3,4), (0,2)(1,3), (0,2)(1,3,4), (0,2,5)(1,3), (0,2,5)(1,3,4), (0,2)(1,4,3), (0,2)(1,4), (0,2,5)(1,4,3), (0,2,5)(1,4), (0,3,2,1)(4,5), (0,3,5,4,2,1), (0,3,2,4,5,1), (0,3,5,1)(2,4), (0,3)(1,2)(4,5), (0,3,5,4)(1,2), (0,3)(1,2,4,5), (0,3,5,1,2,4), (0,3)(1,5,4,2), (0,3,2,1,5,4), (0,3)(1,5)(2,4), (0,3,2,4)(1,5), (0,4,5,3,2,1), (0,4,2,1)(3,5), (0,4,5,1)(2,3), (0,4,2,3,5,1), (0,4,5,3)(1,2), (0,4)(1,2)(3,5), (0,4,5,1,2,3), (0,4)(1,2,3,5), (0,4,2,1,5,3), (0,4)(1,5,3,2), (0,4,2,3)(1,5), (0,4)(1,5)(2,3), (0,5,2), (0,5,2)(3,4), (0,5), (0,5)(3,4), (0,5,2)(1,3), (0,5,2)(1,3,4), (0,5)(1,3), (0,5)(1,3,4), (0,5,2)(1,4,3), (0,5,2)(1,4), (0,5)(1,4,3), (0,5)(1,4)	3-3 hierarchical symmetry
53	S_5	Partial symmetry in 5 variables.

Figure 25. The subgroups of S_6 Part 4.

Group 26 is similar to Group 25, except the swap (4,5) is associated with the generator (1,3) instead of (0,1)(2,3). Group 26 is generated by (0,1,2,3)(4,5) and (1,3)(4,5).

54	I, (2,3)(4,5), (2,4,3,5), (2,5,3,4), (1,2)(3,4), (1,2,3,5), (1,2,4,5,3), (1,2,5,4), (1,3,5,4,2), (1,3,4,5), (1,3,2,4), (1,3)(2,5), (1,4,5,2), (1,4)(3,5), (1,4,2,3), (1,4,3,2,5), (1,5,3,2), (1,5,4,3), (1,5,2,3,4), (1,5)(2,4), (0,1)(4,5), (0,1)(2,3), (0,1)(2,4)(3,5), (0,1)(2,5)(3,4), (0,1,2)(3,5,4), (0,1,2,3,4,5), (0,1,2,4), (0,1,2,5,3), (0,1,3,4,2), (0,1,3,5), (0,1,3)(2,4,5), (0,1,3,2,5,4), (0,1,4,5,3,2), (0,1,4,3), (0,1,4)(2,3,5), (0,1,4,2,5), (0,1,5,2), (0,1,5,3,4), (0,1,5,4,2,3), (0,1,5)(2,4,3), (0,2,1)(3,4,5), (0,2,3,5,4,1), (0,2,4,3,1), (0,2,5,1), (0,2)(3,5), (0,2,3,4), (0,2,4,5), (0,2,5,4,3), (0,2)(1,3)(4,5), (0,2,1,3), (0,2,4)(1,3,5), (0,2,5)(1,3,4), (0,2)(1,4), (0,2,1,4,3,5), (0,2,3)(1,4,5), (0,2,5,3,1,4), (0,2)(1,5)(3,4), (0,2,1,5,4), (0,2,3,1,5), (0,2,4,1,5,3), (0,3,5,2,1), (0,3,4,1), (0,3,2,4,5,1), (0,3,1)(2,5,4), (0,3,4,5,2), (0,3,5,4), (0,3)(2,4), (0,3,2,5), (0,3,1,2), (0,3)(1,2)(4,5), (0,3,5)(1,2,4), (0,3,4)(1,2,5), (0,3,5,1,4,2), (0,3,2,1,4), (0,3,1,4,5), (0,3)(1,4)(2,5), (0,3,2)(1,5,4), (0,3,4,2,1,5), (0,3)(1,5), (0,3,1,5,2,4), (0,4,2,1), (0,4,3,5,1), (0,4,5,2,3,1), (0,4,1)(2,5,3), (0,4,3,2), (0,4,5,3), (0,4,2,3,5), (0,4)(2,5), (0,4,5,1,2), (0,4)(1,2)(3,5), (0,4,1,2,3), (0,4,3,1,2,5), (0,4,1,3,5,2), (0,4,5)(1,3,2), (0,4)(1,3), (0,4,2,5,1,3), (0,4,2)(1,5,3), (0,4,3)(1,5,2), (0,4,1,5), (0,4)(1,5)(2,3), (0,5,4,3,2,1), (0,5,3,1), (0,5,1)(2,3,4), (0,5,2,4,1), (0,5,4,2), (0,5)(3,4), (0,5,2,3), (0,5,3,2,4), (0,5,3,4,1,2), (0,5)(1,2), (0,5,4)(1,2,3), (0,5,1,2,4,3), (0,5,1,3,2), (0,5,2,1,3,4), (0,5,4,1,3), (0,5)(1,3)(2,4), (0,5,2)(1,4,3), (0,5,3)(1,4,2), (0,5,1,4), (0,5)(1,4)(2,3)	Forbidden
55	A_6	Forbidden
56	S_6	Total

Figure 26. The subgroups of S_6 Part 5.

Group 32 is similar to Group 9. A function with this symmetry will have three pairs of variables, any two of which can be swapped simultaneously. In addition, it is possible to select one input from each pair and rotate them, as long as the unselected pairs are rotated in exactly the same way.

Group 41 is an "enhanced" form of 4-variable partial symmetry. The first four variables are partially symmetric, but any odd permutation of the first four variables requires a simultaneous swap of the last two variables.

The symmetry of Group 42 is similar to dihedral symmetry, but more complicated. Arrange the numbers 0 through 5 around the six corners of a hexagon. Any permutation, except reflection, that leaves the pairs (0,3), (1,4), and (2,5) diametrically opposed to one another is acceptable. This is neither dihedral nor rotational symmetry, as an examination of the elements of Group 42 will show. Group 49 is similar to Group 42, except that reflection is permissible.

The symmetry of Group 43 is extremely difficult to describe. It contains four independent symmetries identical to that of Group 17, and three independent symmetries identical to that of Group 13. The independent symmetries are not applied to separate sets of variables, but interlock in a complex way.

Group 51 is an oddity because consists only of even permutations but is not forbidden. (There are others with this property, group 13, for example). The action of this group on a Boolean function is virtually impossible to describe in simple terms.

The other subgroups of S_6 are either forbidden or fall into one of the categories already described.

4. SYMMETRIC VARIABLE PAIRS

A symmetric variable pair (a,b) of a function f is a pair of variables that can be exchanged without altering the output of f. One can detect a symmetric variable pair using the cofactors of f. A cofactor of f is computed by setting one or more variables of f to constant values. Subscripts are used to indicate which variables have been set to constant values, and whether the values are 0 or 1. For example suppose f is the function $x_1x_2 + x_3x_4$. We can create a cofactor of f by setting $x_1 = 0$ and $x_2 = 1$. Thus $f_{01xx} = 0 \cdot 1 + x_3x_4 = 0 + x_3x_4 = x_3x_4$. The subscript indicates the replacement values as well as the variables that have not been replaced.

When the replacement values are obvious, the x's are omitted from the subscript. Shannon's theorem tells us that (a,b) is a symmetric variable pair if and only if $f_{01} = f_{10}$, where the replacements are done with respect to the variables (a,b).

The concept of symmetry can be extended by using cofactor relations beyond the simple $f_{01} = f_{10}$. There are four cofactors with respect to a pair of variables: f_{00}, f_{01}, f_{10}, and f_{11}. There are six possible equality relations between these cofactors, each one of which can define a type of symmetry. Figure 26 lists these equality relations, and the type of symmetry defined by each.

As mentioned in Section 2, multi-phase symmetry is symmetry with respect to functions with mixed active-high/active-low inputs. The test

$f_{00} = f_{11}$ does not tell us which signal is active-high, only that the phases of the two inputs differ, but for proper handling of symmetry, this is all that is required. The single-variable and multi-phase single-variable symmetries are types of conjugate symmetry, which is described in Section 6.

Relation	Symmetry Type
$f_{01} = f_{10}$	Ordinary
$f_{00} = f_{11}$	Multi-Phase
$f_{01} = f_{11}$	Single-Variable
$f_{10} = f_{11}$	Single-Variable
$f_{01} = f_{00}$	Multi-Phase + Single Variable
$f_{10} = f_{00}$	Multi-Phase + Single Variable

Figure 27. Cofactor relations and symmetry types.

The six cofactor relations of Figure 27 can be recast in the following form: $f_{01} \oplus f_{10} = \hat{0}$, where $\oplus$ is exclusive-or and $\hat{0}$ is the constant-zero function. This permits additional cofactor relations to be defined. For example, the relation $f_{01} \oplus f_{10} = \hat{1}$ (where $\hat{1}$ is the constant-one function) defines ordinary anti-symmetry. Figure 28 lists the cofactor relations that define the various types of anti-symmetry. Anti-symmetric variable pairs are just as common as their positive counterparts and can be exploited in much the same way.

Relation	Anti-Symmetry Type
$f_{01} \oplus f_{10} = \hat{1}$	Ordinary
$f_{00} \oplus f_{11} = \hat{1}$	Multi-Phase
$f_{01} \oplus f_{11} = \hat{1}$	Single-Variable
$f_{10} \oplus f_{11} = \hat{1}$	Single-Variable
$f_{01} \oplus f_{00} = \hat{1}$	Multi-Phase + Single Variable
$f_{10} \oplus f_{00} = \hat{1}$	Multi-Phase + Single Variable

Figure 28. Cofactor relations for Anti-Symmetry.

Anti-symmetry can also be defined in terms of Boolean Orbits. Let f be a Boolean function and $G=\{I,(a,b)\}$, where (a,b) is a transposition. If every characteristic function of the Boolean orbits of G implies either f or $\overline{f}$ then (x_a,x_b) is an ordinary symmetric variable pair of f. If every characteristic function implies *neither* f nor $\overline{f}$, then (x_a,x_b) is an ordinary anti-symmetric pair of f. This concept can be extended to arbitrary sets of Boolean orbits, but it's not clear that this concept is useful.

Using the exclusive-or concept, it is possible to extend cofactor relations to three or more cofactors. For example, the relation $f_{00} \oplus f_{01} \oplus f_{10} = \hat{0}$ defines a type of Kronecker symmetry. Figure 29 shows the relations defining the various types of Kronecker symmetry.

Positive and Negative Kronecker Relations	
Positive	Negative
$f_{00} \oplus f_{01} \oplus f_{10} = \hat{0}$	$f_{00} \oplus f_{01} \oplus f_{10} = \hat{1}$
$f_{00} \oplus f_{01} \oplus f_{11} = \hat{0}$	$f_{00} \oplus f_{01} \oplus f_{11} = \hat{1}$
$f_{00} \oplus f_{10} \oplus f_{11} = \hat{0}$	$f_{00} \oplus f_{10} \oplus f_{11} = \hat{1}$
$f_{01} \oplus f_{10} \oplus f_{11} = \hat{0}$	$f_{01} \oplus f_{10} \oplus f_{11} = \hat{1}$
$f_{00} \oplus f_{01} \oplus f_{10} \oplus f_{11} = \hat{0}$	$f_{00} \oplus f_{01} \oplus f_{10} \oplus f_{11} = \hat{1}$

Figure 29. The Kronecker Relations.

Kronecker symmetries are quite common, but more research is needed to determine how to exploit them most effectively.

Of course cofactor relations can be extended further in a number of different ways. The exclusive-or operator could be replaced by functions such as AND, OR, or something else. The right-hand side could be replaced by some non-constant function.

These extensions would allow one to define many new types of symmetry, but as yet it is not clear how one would exploit these types of symmetry. Clearly, a great deal of research is needed in this area.

5. MATRIX-BASED SYMMETRY

5.1. Conjugate Symmetry

Another mechanism that can be used to extend the concept of symmetry to new functions is the idea of *conjugate symmetry.* Conjugate symmetry involves a change of variable where the existing variables of a function are replaced with certain linear combinations of the variables. If the function is symmetric with respect to these new variables then it is said to possess conjugate symmetry. At the present time there are few algorithms that can take advantage of this type of symmetry, but it is enormously beneficial to those algorithms that can take advantage of it. Conjugate symmetry is far more common than ordinary symmetry, so much so that most Boolean functions used in practice exhibit some form of conjugate symmetry.

To fully explain conjugate symmetry it is necessary to formalize some of the intuitive ideas that we have been using up to this point. Specifically, we must formalize the concept of applying a permutation to a Boolean vector.

Up to this point we have visualized a Boolean vector as a sequence of zeros and ones of the form $V=(v_0,v_1,...,v_n)$, and we have applied permutations to V by permuting the subscripts $0,1,...,n$. We have ignored the fact that a permutation of the indices is, in fact, a linear transformation of the vector V. To understand conjugate symmetry, it is necessary to focus on these linear transformations. Unfortunately, Boolean algebra, with its operations AND, OR, and NOT, does not possess the necessary algebraic properties to construct linear transformations. Therefore, it is necessary to alter our perspective slightly and consider the algebraic properties of $GF(2)$, the integers modulo 2. Like ordinary Boolean algebra, $GF(2)$ contains two values, zero and one. In addition, $GF(2)$ has two algebraic operations, multiplication and addition. Multiplication corresponds to the Boolean AND operator: one times one is one, and zero times anything is zero. Addition, however corresponds to the Boolean operator Exclusive-Or, XOR: zero plus zero is zero, zero plus one is one, but one plus one is two which corresponds to zero.

If we use only the operations AND and XOR, $GF(2)$ possesses all the algebraic properties we need to construct linear transformations. We can represent linear transformations as matrices, just as we would if we were dealing with real numbers. However, in this case the matrix will contain only the values zero and one. Virtually any operation that can be done on a real matrix can also be done on a $GF(2)$ matrix. We can multiply matrices, add them, perform Gaussian elimination, find characteristic polynomials and eigen values, and so forth.

Every permutation can be represented as a matrix called a *permutation* matrix. A permutation matrix contains a single one in each row and column, and zeros elsewhere. We can represent the group S_n and its subgroups as collections of $n \times n$ matrices. Figure 30 shows how S_3 can be represented as a collection of 3×3 matrices. Figure 30 also shows how each matrix corresponds to a permutation. (The first row of each matrix is row zero.)

$$\underset{I}{\begin{pmatrix}1&0&0\\0&1&0\\0&0&1\end{pmatrix}}\underset{(0,1)}{\begin{pmatrix}0&1&0\\1&0&0\\0&0&1\end{pmatrix}}\underset{(0,2)}{\begin{pmatrix}0&0&1\\0&1&0\\1&0&0\end{pmatrix}}\underset{(1,2)}{\begin{pmatrix}1&0&0\\0&0&1\\0&1&0\end{pmatrix}}\underset{(0,1,2)}{\begin{pmatrix}0&1&0\\0&0&1\\1&0&0\end{pmatrix}}\underset{(0,2,1)}{\begin{pmatrix}0&0&1\\1&0&0\\0&1&0\end{pmatrix}}$$

Figure 30. Permutation Matrices.

To show how this works, Figure 31 shows the application of the matrix for (0,1,2) to the vector $V = (v_0, v_1, v_2)$. This is done using ordinary matrix multiplication, with AND substituted for multiplication and XOR substituted for addition.

Note how this permutes the elements of V.

$$(v_0, v_1, v_2)\begin{pmatrix}0&1&0\\0&0&1\\1&0&0\end{pmatrix} =$$

$$(0\times v_0 + 0\times v_1 + 1\times v_2, 1\times v_0 + 0\times v_1 + 0\times v_2, 0\times v_0 + 1\times v_1 + 0\times v_2) = (v_2, v_0, v_1)$$

Figure 31. Permuting a vector using a matrix.

Let f be an n-input Boolean function and let M be an $n \times n$ matrix. The function f is said to be *invariant* with respect to M if $f(M(v)) = f(v)$. In other words, the composition of f and M is identical to f. An n-input

Boolean function is totally symmetric if and only if it is invariant with respect to all $n \times n$ permutation matrices. Functions that are partially and strongly symmetric are invariant with respect to some subgroup of the permutation matrices.

The set of $n \times n$ permutation matrices is called a *representation* of S_n. Thus the set of matrices given in Figure 26 is called a *representation* of S_3. Because all symmetries can be described in terms of permutation matrices, the representation of S_n consisting of all permutation matrices defines every type of known symmetry. This representation is called the *standard representation* of S_n. We designate the standard representation of S_n as SR_n.

However, the standard representation of S_n is not the *only* representation of S_n. Technically speaking, any homomorphism of S_n into a set of matrices is a representation of S_n. If the homomorphism is an isomorphism, then the representation is called a *faithful representation*. Faithful representations preserve all of the mathematical properties of S_n. There are many faithful representations of S_n. For example Figure 32 gives an alternative representation of S_3. This alternative representation is just as "legitimate" as the standard representation. It defines a complete collection of symmetries, containing total, partial and strong symmetries.

$$\underset{I}{\begin{pmatrix}1&0&0\\0&1&0\\0&0&1\end{pmatrix}}\underset{(0,1)}{\begin{pmatrix}1&1&1\\0&1&0\\0&0&1\end{pmatrix}}\underset{(0,2)}{\begin{pmatrix}1&0&0\\1&1&1\\0&0&1\end{pmatrix}}\underset{(1,2)}{\begin{pmatrix}0&1&0\\1&0&0\\0&0&1\end{pmatrix}}\underset{(0,1,2)}{\begin{pmatrix}0&1&0\\1&1&1\\0&0&1\end{pmatrix}}\underset{(0,2,1)}{\begin{pmatrix}1&1&1\\1&0&0\\0&0&1\end{pmatrix}}$$

Figure 32. An alternative representation of S_3.

While it is true that the representation of Figure 32 does not correspond to our intuitive idea of swapping input variables, the matrices of the representation allow us to define symmetry with respect to this representation in precisely the same manner as was done for the standard representation. That is, let f be a 3-input Boolean function, and let M be a any matrix given in

Figure 32. The function f is said to be invariant with respect to M if and only if $f(M(v)) = f(v)$.

There are a total of 28 different representations of S_3 as a collection of 3×3 matrices, including the standard representation. Each of these defines a complete set of symmetries with total, partial, and strong symmetries. The Boolean orbits of the alternative representations differ from those of the standard representation, however. Figure 33 gives the Boolean orbits of the standard representation, while Figure 34 gives the Boolean orbits of the representation given in Figure 32.

000
001 100 010
011 110 101
111

Figure 33. The Boolean orbits of the standard representation.

000
010 111 100
110 101 011
001

Figure 34 The Boolean orbits of an alternative representation.

There is a strong relationship between the standard representation of S_3 and the representation of Figure 32 Figure 35 gives a non-singular 3×3 matrix followed by its inverse. This matrix can be used to compute the representation of Figure 32 from the standard representation. To do this we use the concept of similar matrices, which is identical to the concept of conjugate permutations presented in Section 2.

$$\begin{pmatrix} 0 & 1 & 0 \\ 0 & 0 & 1 \\ 1 & 1 & 1 \end{pmatrix} \begin{pmatrix} 1 & 1 & 1 \\ 1 & 0 & 0 \\ 0 & 1 & 0 \end{pmatrix}$$

Figure 35. A matrix and its inverse.

Two matrices, M_1 and M_2 are said to be similar (or conjugate to one another) if there is a non-singular matrix T such that $M_1 = TMT^{-1}$. Let G be a group of matrices. Then the conjugate of G with respect to a nonsingular matrix T is the set of matrices $\{TMT^{-1} \mid M \in G\}$. We write TGT^{-1} for the conjugate of G with respect to T. Let T be the matrix of Figure 34. The representation of Figure 32 is $T\,SR_n\,T^{-1}$. Figure 36 shows how the representation of Figure 32 was computed.

$$\begin{pmatrix}0&1&0\\0&0&1\\1&1&1\end{pmatrix}\begin{pmatrix}1&0&0\\0&1&0\\0&0&1\end{pmatrix}\begin{pmatrix}1&1&1\\1&0&0\\0&1&0\end{pmatrix}=\begin{pmatrix}1&0&0\\0&1&0\\0&0&1\end{pmatrix}$$

$$\begin{pmatrix}0&1&0\\0&0&1\\1&1&1\end{pmatrix}\begin{pmatrix}0&1&0\\1&0&0\\0&0&1\end{pmatrix}\begin{pmatrix}1&1&1\\1&0&0\\0&1&0\end{pmatrix}=\begin{pmatrix}1&1&1\\0&1&0\\0&0&1\end{pmatrix}$$

$$\begin{pmatrix}0&1&0\\0&0&1\\1&1&1\end{pmatrix}\begin{pmatrix}0&0&1\\0&1&0\\1&0&0\end{pmatrix}\begin{pmatrix}1&1&1\\1&0&0\\0&1&0\end{pmatrix}=\begin{pmatrix}1&0&0\\1&1&1\\0&0&1\end{pmatrix}$$

$$\begin{pmatrix}0&1&0\\0&0&1\\1&1&1\end{pmatrix}\begin{pmatrix}1&0&0\\0&0&1\\0&1&0\end{pmatrix}\begin{pmatrix}1&1&1\\1&0&0\\0&1&0\end{pmatrix}=\begin{pmatrix}0&1&0\\1&0&0\\0&0&1\end{pmatrix}$$

$$\begin{pmatrix}0&1&0\\0&0&1\\1&1&1\end{pmatrix}\begin{pmatrix}0&1&0\\0&0&1\\1&0&0\end{pmatrix}\begin{pmatrix}1&1&1\\1&0&0\\0&1&0\end{pmatrix}=\begin{pmatrix}0&1&0\\1&1&1\\0&0&1\end{pmatrix}$$

$$\begin{pmatrix}0&1&0\\0&0&1\\1&1&1\end{pmatrix}\begin{pmatrix}0&0&1\\1&0&0\\0&1&0\end{pmatrix}\begin{pmatrix}1&1&1\\1&0&0\\0&1&0\end{pmatrix}=\begin{pmatrix}1&1&1\\1&0&0\\0&0&1\end{pmatrix}$$

Figure 36. Computing a conjugate symmetry.

Let G be a representation of S_n that is conjugate to the standard representation SR_n. Any Boolean function f that is invariant with respect to some element of G other than the identity element, is said to possess *conjugate symmetry*. Normally we identify the conjugacy matrix T, where $G = T\,SR_n\,T^{-1}$, by saying that f is *symmetric with respect to* T.

Although conjugate representations have different Boolean orbits from the standard representation, the orbit *structure* will be identical. For an *n*-input totally symmetric function, there will be *n*+1 Boolean orbits numbered 0 through *n*. Orbit 0 will contain the zero vector and nothing else. Orbit 1 will contain *n* linearly independent vectors. Orbit k will contain $\binom{n}{k}$ vectors, and will be the collection of all possible sums of k vectors chosen from Orbit 1.

Functions that are symmetric with respect to a non-singular matrix T can be factored into a non-singular matrix and an ordinary symmetric function. Thus if f is symmetric with respect to T, then $f(v) = g(M(v))$ where M is a non-singular matrix, and g is an ordinary symmetric function. If f is totally symmetric with respect to T, then g will be totally symmetric. If f is partially symmetric with respect to T then g will be partially symmetric in exactly the same way. The same holds true for the various types of strong symmetry.

One obvious factorization can be obtained by setting $M = T^{-1}$, although several different factorizations are possible. To see why this is so, let g be a symmetric function that is invariant with respect to $N \in SR_n$. If $K = TNT^{-1}$ then $g(T^{-1}(v))$ will be invariant with respect to K, since $g(T^{-1}(TNT^{-1}(v))) = g(NT^{-1}(v)) = g(T^{-1}(v))$.

Partial and total conjugate symmetries can be detected using symmetric variable pairs. This is done by mapping the function values onto the elements of a 2^n-element lattice which has the form of an *n*-dimensional hypercube. The conjugate-symmetry variable pairs can be detected by analyzing the effect of a linear transformation on the lattice.

Every non-singular matrix can be decomposed into a product of elementary matrices. In GF(2), an elementary matrix is reduced to a matrix with ones along the main diagonal, a single one in a position off the main diagonal, and zeros elsewhere. The effect of applying an elementary matrix to the lattice of n-bit values is to reflect the odd-numbered planes of the lattice in one particular direction. The symmetric variable pair can be detected by reflecting the odd planes in a single direction and then testing for ordinary (or other) symmetry. The conjugacy matrix is accumulated as a product of elementary matrices, and the alterations to the lattice produce the desired symmetric function.

The entire process is complex and is fully described in [3].

5.2. Super Symmetry

The most natural representations of S_n tend to be those where S_n is mapped into a set of $n \times n$ matrices, however it is possible to map S_n into a set of $m \times m$ matrices where $m \neq n$. In fact, certain partial symmetries can be defined as mappings from S_n to a set of $m \times m$ matrices where $m > n$. It is also possible to map S_n into a set of $(n-1)\times(n-1)$ matrices using the following procedure.

1. Let v_i be the vector that contains a 1 in position i and zeros elsewhere.
2. Let v_0 be the vector containing a 1 in each position.
3. Let SH_n be the set of all non-singular matrices that can be formed from the rows v_0 through v_n.
4. Let SV_n be the set of all non-singular matrices that can be formed by the columns v_0 through v_n.

Both SH_n and SV_n are isomorphic to S_{n+1}. Any n-input function that is invariant with respect to SH_n or SV_n is said to be *super-symmetric*. Since the standard representation of S_n is a subset of both SH_n and SV_n any super-symmetric function will also be totally symmetric. Figure 37 shows the matrix representation of SH_3 and Figure 38 shows the matrix representation of SV_3.

It is extremely easy to analyze the structure of the Boolean orbits of SH_n and SV_n.

$$\begin{pmatrix}1&0&0\\0&1&0\\0&0&1\end{pmatrix}\begin{pmatrix}0&1&0\\1&0&0\\0&0&1\end{pmatrix}\begin{pmatrix}0&0&1\\0&1&0\\1&0&0\end{pmatrix}\begin{pmatrix}1&0&0\\0&0&1\\0&1&0\end{pmatrix}\begin{pmatrix}0&1&0\\0&0&1\\1&0&0\end{pmatrix}\begin{pmatrix}0&0&1\\1&0&0\\0&1&0\end{pmatrix}$$

$$\begin{pmatrix}1&1&1\\0&1&0\\0&0&1\end{pmatrix}\begin{pmatrix}1&1&1\\1&0&0\\0&0&1\end{pmatrix}\begin{pmatrix}1&1&1\\0&1&0\\1&0&0\end{pmatrix}\begin{pmatrix}1&1&1\\0&0&1\\0&1&0\end{pmatrix}\begin{pmatrix}1&1&1\\0&0&1\\1&0&0\end{pmatrix}\begin{pmatrix}1&1&1\\1&0&0\\0&1&0\end{pmatrix}$$

$$\begin{pmatrix}1&0&0\\1&1&1\\0&0&1\end{pmatrix}\begin{pmatrix}0&1&0\\1&1&1\\0&0&1\end{pmatrix}\begin{pmatrix}0&0&1\\1&1&1\\1&0&0\end{pmatrix}\begin{pmatrix}1&0&0\\1&1&1\\0&1&0\end{pmatrix}\begin{pmatrix}0&1&0\\1&1&1\\1&0&0\end{pmatrix}\begin{pmatrix}0&0&1\\1&1&1\\0&1&0\end{pmatrix}$$

$$\begin{pmatrix}1&0&0\\0&1&0\\1&1&1\end{pmatrix}\begin{pmatrix}0&1&0\\1&0&0\\1&1&1\end{pmatrix}\begin{pmatrix}0&0&1\\0&1&0\\1&1&1\end{pmatrix}\begin{pmatrix}1&0&0\\0&0&1\\1&1&1\end{pmatrix}\begin{pmatrix}0&1&0\\0&0&1\\1&1&1\end{pmatrix}\begin{pmatrix}0&0&1\\1&0&0\\1&1&1\end{pmatrix}$$

Figure 37. The matrix representation of SH_3.

$$\begin{pmatrix}1&0&0\\0&1&0\\0&0&1\end{pmatrix}\begin{pmatrix}0&1&0\\1&0&0\\0&0&1\end{pmatrix}\begin{pmatrix}0&0&1\\0&1&0\\1&0&0\end{pmatrix}\begin{pmatrix}1&0&0\\0&0&1\\0&1&0\end{pmatrix}\begin{pmatrix}0&1&0\\0&0&1\\1&0&0\end{pmatrix}\begin{pmatrix}0&0&1\\1&0&0\\0&1&0\end{pmatrix}$$

$$\begin{pmatrix}1&0&0\\1&1&0\\1&0&1\end{pmatrix}\begin{pmatrix}1&1&0\\1&0&0\\1&0&1\end{pmatrix}\begin{pmatrix}1&0&1\\1&1&0\\1&0&0\end{pmatrix}\begin{pmatrix}1&0&0\\1&0&1\\1&1&0\end{pmatrix}\begin{pmatrix}1&1&0\\1&0&1\\1&0&0\end{pmatrix}\begin{pmatrix}1&0&1\\1&0&0\\1&1&0\end{pmatrix}$$

$$\begin{pmatrix}1&1&0\\0&1&0\\0&1&1\end{pmatrix}\begin{pmatrix}0&1&0\\1&1&0\\0&1&1\end{pmatrix}\begin{pmatrix}0&1&1\\0&1&0\\1&1&0\end{pmatrix}\begin{pmatrix}1&1&0\\0&1&1\\0&1&0\end{pmatrix}\begin{pmatrix}0&1&0\\0&1&1\\1&1&0\end{pmatrix}\begin{pmatrix}0&1&1\\1&1&0\\0&1&0\end{pmatrix}$$

$$\begin{pmatrix}1&0&1\\0&1&1\\0&0&1\end{pmatrix}\begin{pmatrix}0&1&1\\1&0&1\\0&0&1\end{pmatrix}\begin{pmatrix}0&0&1\\0&1&1\\1&0&1\end{pmatrix}\begin{pmatrix}1&0&1\\0&0&1\\0&1&1\end{pmatrix}\begin{pmatrix}0&1&1\\0&0&1\\1&0&1\end{pmatrix}\begin{pmatrix}0&0&1\\1&0&1\\0&1&1\end{pmatrix}$$

Figure 38. The matrix representation of SV_3.

First we start with the Boolean orbits of the standard representation of S_n, and number them from 0 through n, depending on the weight (the number of 1's in the vector) of the vectors in the orbit. The significant transformations in SH_n are those in which a row of the identity transformation is replaced with a row of all ones. Similarly, the significant transformations of SV_n are those in which a column of the identity matrix is replaced with a column of all ones. The other new transformations in SH_n and SV_n are merely a significant transformation followed by a permutation.

When a significant transformation of SH_n is applied to a vector of weight k, one of the ones in the vector is held constant, while the other $n-1$ bits are inverted. The weight of the affected bits is $k-1$. Therefore the weight of the affected bits after transformation is $(n-1)-(k-1)=n-k$ and the weight of the new vector is $n-k+1$. Thus the significant transformations have the effect of combining orbit k with orbit $n-k+1$. For 4 inputs this combines orbit 1 with orbit 4-1+1=4, and orbit 2 with orbit 4-2+1=3.

For 5 inputs this combines orbit 1 with orbit 5-1+1=5, orbit 2 with orbit 5-2+1=4 and orbit 3 with 5-3+1=3. (Orbit 3 is unaffected.)

When a significant transformation of SV_n is applied to a vector $v=(v_1,v_2,\ldots,v_n)$, it replaces one of the bits of the vector with the weight of the vector modulo 2: $v_1+v_2+\cdots+v_n$. This has the effect of permuting the $n+1$ values v_1, v_2, ... , v_n, and $v_1+v_2+\cdots+v_n$, one of which will be "missing" from the vector.

Suppose k is an odd-numbered orbit of S_n. The modulo 2 weight of the vectors in this orbit will be 1. We can replace any bit of a vector with this weight. If we replace a 1 the weight of the vector does not change and we obtain another vector from orbit k. If we replace a 0, the weight will increase by 1 and the vector will be a member of orbit $k+1$. Thus the significant transformations have the effect of combining the odd-numbered orbits with the next highest numbered orbit. If $n=4$ then SH_n and SV_n will have the same Boolean orbits, (and are conjugate to one another). If $n=2$ then $SH_n=SV_n$. For all other values of n, SH_n and SV_n give different types of symmetry.

Figure 39 gives the Boolean orbits of SH_3 and gives a pair of super-symmetric functions along with their Karnaugh maps.

000
100, 010, 001, 111
011, 101, 110

a/bc	00	01	11	10
0		1		1
1	1		1	

$ab'c' + a'bc' + a'b'c + abc$

a/bc	00	01	11	10
0	1	1		1
1	1		1	

$a'b' + a'c' + b'c' + abc$

Figure 39. Two Super-symmetric functions.

Of course, the matrix groups of Figures 37 and 38 tell only part of the story. To be more specific, these matrix groups define *total* super-symmetry. The subgroups of these matrix groups define partial super-symmetry and strong super-symmetry. Input-inversions can be used to define multi-phase super-symmetry just as it was defined in Section 4. Furthermore, the techniques of Section 5.1 can be exploited to define entire new classes of super-symmetry. Each of these classes of super-symmetry contain a complete set of total super-symmetries, partial super-symmetries, strong super-symmetries and multi-phase super-symmetries. Although the possibilities are intriguing, little is known about these symmetry classes.

5.3. Exotic Symmetry

As noted in Section 5.1, any representation of S_n in $n \times n$ matrices gives a new class of symmetries that is identical in structure to ordinary symmetry. Section 5.1 also explored the nature of representations that are conjugate to the standard representation, but for $n > 3$, there are representations of S_n that are *not* conjugate to the standard representation. We call these symmetries *exotic symmetries*.

Taken together, the representations that are conjugate to the standard representation form a superclass of symmetries that all possess the same Boolean-orbit structure. Exotic symmetries belong to their own superclasses of symmetries with "standard" representations and collections of additional representations that are conjugate to the standard representation of the class.

Because these representations do not conform to our intuitive idea of permuting inputs (or linear combinations of inputs) it is difficult to pick one representation as "standard." Nevertheless, we have endeavored to do so by picking a representation that conforms as closely as possible to the Boolean orbits of the standard representation of ordinary symmetry.

There are many open problems in exotic symmetries, two most important of which are enumerating the different superclasses, and determining how to utilize them effectively. At this time we have been able to enumerate the superclasses for 4 and 5-input functions only. This is due to the enormous computational complexity of the algorithms used to enumerate the superclasses for 4 and 5 inputs. There are 20,160 non-singular 4x4 matrices, making the enumeration problem comparatively easy. However, there are over nine million 5x5 non-singular matrices, making the enumeration problem just barely feasible on an ordinary computer. For 6x6 matrices, the total is over 20 billion, making the problem out of reach by existing methods. There are mathematical techniques that could be applied, but not enough information is currently available to be able to apply these techniques with any certainty.

For 4-input functions, there are nine superclasses, including the superclass containing ordinary symmetry. For 5-input functions there are only four such superclasses. A few superclasses have been identified for 6-input functions, but this work is far from complete.

Figure 40 gives the structure of the Boolean orbits of the 4-input superclasses and shows how they relate to the orbital structure of the standard representation of ordinary symmetry. We designate ordinary symmetry as Superclass 0, and the other superclasses as Superclass A through Superclass H. These letters bear no special significance. They conform to the order in which the superclasses were discovered.

In addition to the nine superclasses described above, there is an exotic super-symmetry which is not conjugate to either SH_4 or SV_4. We call this Super-symmetry-A in keeping with the notation of Figure 40.

Superclass	Boolean Orbits	Structure
0	0000 0001 0010 0100 1000 0011 0101 1001 0110 1010 1100 0111 1011 1101 1110 1111	Orbit 1 contains 4 linearly independent vectors. Orbits 2 through 4 contain linear combinations of 2, 3, and 4 vectors from Orbit 1.
A	0000 0001 0010 1100 0100 1001 1110 1000 0011 1101 1011 0110 0111 0101 1010 1111	Splits out 0101 and 1010 from Orbit 2 of the standard representation. Orbit 3 is the sum of the vectors in orbit 2. All other vectors are placed in Orbit 1.
B	0000 0001 1011 0010 1000 0100 0111 1110 1101 0011 1100 0110 0101 1001 1010 1111	Combines Orbits 1 and 3 from the standard representation.
C	0000 0001 1101 0010 1000 0100 1011 0111 1110 0011 0110 1100 1001 0101 1111 1010	Combines Orbits 1 and 3 from the standard representation. Orbit 2 is split with 0101 and 1010 being removed and combined with Orbit 4. All orbits sum to zero.
D	0000 0001 0010 0100 0111 1100 1001 1110 1101 1011 1000 0011 0110 0101 1010 1111	Orbits 1 and 2 are complements of one another, as are orbits 3 and 4.
E	0000 0001 1110 0010 1101 0110 1001 0011 1000 1011 0100 0111 1100 0101 1010 1111	Similar to D with orbits 3 and 4 combined.
F	0000 0001 1101 0100 1000 1011 1110 0111 0010 0011 1010 0110 1100 0101 1001 1111	Combines Orbits 1 and 3, and splits Orbit 2 into two orbits which are the complements of one another.
G	0000 0001 0100 0010 0111 1110 1011 1101 1000 0011 0110 0101 1100 1001 1010 1111	Orbits 1 and 2 are complements of one another, as are orbits 3 and 4.
H	0000 0001 1001 0010 0100 0011 1101 1000 0110 1011 1100 1110 0111 0101 1010 1111	Combines Orbits 1, 2 and 3, except for 0101 and 1010 which are combined with orbit 4.

Figure 40. The Superclasses of 4-input symmetry.

The Boolean orbits of Super-symmetry-A are trivial, with orbits 1, 2, 3, and 4 of the standard representation being combined into a single orbit. The AND and NAND functions are the only non-trivial functions possessing Super-symmetry-A. However, this superclass may be important because of its partial and strong symmetries.

5.4. Hyper Symmetry

They final type of matrix symmetry is hyper-symmetry. Hyper-symmetry includes every type of matrix-based symmetry described above. Let G be any group of $n \times n$ matrices. If an n-input function, f, is invariant with respect to every element of G, it is said to have hyper-symmetry of type G. Figure 41 gives three types of symmetry that are not included in any of the symmetry types described above. Each of these symmetries are generated by a single matrix, which is given in Figure 37.

Matrix	Order	Boolean Orbits
$\begin{pmatrix} 0 & 0 & 0 & 1 \\ 0 & 0 & 1 & 0 \\ 0 & 1 & 0 & 1 \\ 1 & 0 & 1 & 1 \end{pmatrix}$	5	0000 0001 1011 1111 1101 1000 0010 0101 1001 1010 0100 0011 1110 0110 0111 1100
$\begin{pmatrix} 0 & 0 & 0 & 1 \\ 0 & 0 & 1 & 0 \\ 0 & 1 & 0 & 1 \\ 1 & 0 & 1 & 0 \end{pmatrix}$	6	0000 0001 1010 0100 0010 0101 1000 0011 1111 1100 0110 0111 1101 1001 1011 1110
$\begin{pmatrix} 1 & 0 & 0 & 1 \\ 0 & 0 & 0 & 1 \\ 0 & 1 & 0 & 0 \\ 0 & 0 & 1 & 1 \end{pmatrix}$	7	0000 0001 0010 0100 0011 0110 0101 0111 1000 1001 1010 1101 1011 1110 1100 1111

Figure 41. Some Hyper-symmetries.

To form the symmetry groups that generate the Boolean orbits of Figure 41, it is necessary to compute all powers of each matrix. (The order i of the matrix M is the smallest power for which $M^i = I$, the identity matrix.) These groups, however, are not necessarily the full symmetry group for the given set of Boolean orbits. In some cases, these are forbidden subgroups of larger hyper-symmetric groups. In fact, the standard representations of several of the superclasses given in Figure 40 are actually forbidden subgroups of larger hyper-symmetric classes. The groups of superclasses A, B, and C are forbidden subgroups of hyper-symmetric groups of size 192. Those of superclasses E and F are forbidden subgroups of hyper-symmetric groups of size 48, and those of superclass H are forbidden subgroups of hyper-symmetric groups of size 576.

Virtually nothing is known about these or any other hyper-symmetric groups. Clearly much work is needed to determine whether these types of symmetry are of value.

Conclusion

Symmetry detection and the enumeration of various different types of symmetry has come a long way since Shannon's initial work. The number of known symmetry types has expanded prodigiously, and the number algorithms to detect these types of symmetry have expanded as well. However, there are many types of symmetry that have yet to be thoroughly investigated and exploited. Indeed, one is left with the feeling, that despite the huge amount of progress that has been made in the area, we have barely scratched the surface of symmetric Boolean functions and their applications. In the upcoming years, we are bound to see considerably more progress in this area.

References

[1] Shannon, C. E., "The Synthesis of Two-Terminal Switching Circuits," The Bell System Technical Journal, Vol. 28, pp. 59-98, 1949.

[2] D. F. Holt. Enumerating subgroups of the symmetric group. *Computational Group Theory and the Theory of Groups, II* pp. 33-37. 2010.

[3] P. M. Maurer, "Conjugate Symmetry," *Formal Methods Syst. Des.,* vol. 38, pp. 263-288, 2011.

BIBLIOGRAPHY

A. Abdollahi, "Canonical form based Boolean matching and symmetry detection in logic synthesis and verification," 2006.

N. N. Biswas, "On Identification of Totally Symmetric Boolean Functions," *Computers, IEEE Transactions On,* vol. 19, pp. 645-648, 1970.

R. C. Born and A. K. Scidmore, "Transformation of switching functions to completely symmetric switching functions," *IEEE Transactions on Computers,* vol. 17, pp. 596-599, 1968.

J. T. Butler, G. W. Dueck, V. P. Shmerko and S. Yanuskevich, "Comments on "Sympathy: fast exact minimization of fixed polarity Reed-Muller expansion for symmetric functions"," *IEEE Transactions on Computer-Aided Design of Integrated Circuits and Systems,* vol. 19, pp. 1386-1388, 2000.

M. Chrzanowska-Jeske, "Generalized symmetric and generalized pseudo-symmetric functions," in *Proceedings of the 6th IEEE International Conference on Electronics, Circuits and Systems,* 1999, pp. 343-346.

M. Chrzanowska-Jeske, "Generalized symmetric variables," in *The 8th IEEE International Conference on Electronics, Circuits and Systems,* 2001, pp. 1147-1150.

K. S. Chung and C. L. Liu, "Local transformation techniques for multi-level logic circuits utilizing circuit symmetries for power reduction," in *Proceedings of the 1998 International Symposium on Low Power Electronics and Design,* 1998, pp. 215-220.

P. T. Darga, K. A. Sakallah and I. L. Markov, "Faster symmetry discovery using sparsity of symmetries," in *Proceedings of the 45th Annual Design Automation Conference,* 2008, pp. 149-154.

R. Drechsler and B. Becker, "Sympathy: Fast exact minimization of fixed polarity reed-muller expressions for symmetric functions," in *European Design and Test Conference,* 1995, pp. 91-97.

D. F. Holt. Enumerating subgroups of the symmetric group. *Computational Group Theory and the Theory of Groups, II* pp. 33-37. 2010.

B. Hu and M. Marek-sadowska, "In-place delay constrained power optimization using functional symmetries," in *Design Automation and Test in Europe,* 2001, pp. 377-382.

Y. Hu, V. Shih, R. Majumdar and L. He, "Exploiting Symmetries to Speed Up SAT-Based Boolean Matching for Logic Synthesis of FPGAs," *IEEE Transactions on Computer-Aided Design of Integrated Circuits and Systems,* vol. 27, pp. 1751-1760, 2008.

W. Ke and P. R. Menon, "Delay-testable implementations of symmetric functions," *IEEE Transactions on Computer-Aided Design of Integrated Circuits and Systems,* vol. 14, pp. 772-775, 1995.

N. Kettle and A. King, "An anytime algorithm for generalized symmetry detection in ROBDDs," *IEEE Transactions on Computer-Aided Design of Integrated Circuits and Systems,* vol. 27, pp. 764-777, 2008.

V. N. Kravets and K. A. Sakallah, "Generalized symmetries in boolean functions," in *IEEE International Conference on Computer Aided Design,* 2000, pp. 526-532.

V. N. Kravets and K. A. Sakallah, "Generalized symmetries in boolean functions," Advanced Computer Architecture Laboratory Department of Electrical Engineering and Computer Science, The University of Michigan, Ann Arbor, MI 48109, 2002.

P. M. Maurer, "Conjugate Symmetry," *Formal Methods Syst. Des.,* vol. 38, pp. 263-288, 2011.

J. Mohnke, P. Molitor and S. Malik, "Limits of using signatures for permutation independent Boolean comparison," *Formal Methods Syst. Des.,* vol. 21, pp. 167-191, 2002.

D. Moller, J. Mohnke and M. Weber, "Detection of symmetry of boolean functions represented by ROBDDs," in *IEEE International Conference on Computer-Aided Design,* 1993, pp. 680-684.

J. C. Muzio, D. M. Miller and S. L. Hurst, "Multivariable symmetries and their detection," *IEE Proceedings on Computers and Digital Techniques,* vol. 130, pp. 141-148, 2008.

D. S. Passman, *Permutation Groups.* New York: W. A. Benjamin, 1968.

G. Pfeiffer, *Subgroups of alternating and symmetric groups,* http://schmidt.nuigalway.ie/subgroups.

J. Rice and J. Muzio, "Antisymmetries in the realization of boolean functions," in *IEEE International Symposium on Circuits and Systems,* 2002, pp. 69-72.

D. Robinson, *A Course in the Theory of Groups.* New York: Springer, 1995.

C. Scholl, S. Melchior, G. Hotz and P. Molitor, "Minimizing ROBDD sizes of incompletely specified boolean functions by exploiting strong symmetries," in *European Design and Test Conference,* 1997, pp. 229-234.

C. E. Shannon, "The synthesis of two-terminal switching circuits," *Bell System Technical Journal,* vol. 28, pp. 59-98, 1949.

C. C. Tsai and M. Marek-Sadowska, "Boolean matching using generalized reed-muller forms," in *Proceedings of the 31st Annual Design Automation Conference,* 1994, pp. 339-344.

C. C. Tsai and M. Marek-Sadowska, "Generalized Reed-Muller forms as a tool to detect symmetries," *IEEE Transactions on Computers,* vol. 45, pp. 33-40, 1996.

K. H. Wang and J. H. Chen, "Symmetry detection for incompletely specified functions," in *Proceedings of the 41st Annual Design Automation Conference,* 2004, pp. 434-437.

J. S. Zhang, M. Chrzanowska-Jeske, A. Mishchenko and J. R. Burch, "Generalized symmetries in boolean functions: Fast computation and application to boolean matching," in *International Workshop on Logic Synthesis,* 2004, pp. 424-430.

In: Boolean Functions
Editor: Allen Hines
ISBN: 978-1-63483-023-2

Chapter 2

BOOLEAN FUNCTIONS: ALL-OPTICAL IMPLEMENTATION USING QUANTUM-DOT SEMICONDUCTOR OPTICAL AMPLIFIERS IN MACH-ZEHNDER INTERFEROMETER

K. E. Zoiros*, E. Dimitriadou and T. Houbavlis

Democritus University of Thrace,
Department of Electrical and Computer Engineering,
Lightwave Communications Research Group, Xanthi, Greece

ABSTRACT

Boolean functions provide the necessary framework for expressing the operation of logic gates, which are the key building units for the accomplishment of signal processing tasks in fundamental and system-oriented level. With the unceasing increase of digital information volume, the capability of executing these functions has been progressively evolved from electronics to photonics so as to avoid the severe constraints imposed in the former domain. This book chapter reviews how this can be done exclusively by means of light by exploring the novel technology of quantum-dot semiconductor optical amplifiers (QD-SOAs), which are incorporated in a switching structure formed by the

* Corresponding author: Tel: +30 25410 79 975, Fax: +30 25410 79 595, Email: kzoiros@ee.duth.gr.

Mach-Zehnder interferometer (MZI). For this purpose we initially present some state-of-the-art technologies that are available for realizing all-optically logic operations, with particular emphasis on the well-established option of SOAs. We underline the limitations in handling ultrafast information when using conventional-type SOAs and we comment on the various solutions that have been proposed for overcoming them. We highlight the need for adopting the novel approach of QD-SOAs in order to successfully confront the challenge of keeping pace with the rapid increase of single channel data rates. We justify this choice by referring to the attractive inherent properties of these devices and how they are combined with those of the MZI to achieve ultra high speed switching. We physically and theoretically describe the operation of QD-SOAs, first as stand-alone entities and then when they are incorporated in the MZI. We demonstrate how the basic configuration of the latter can be modified and driven so as to produce different Boolean functions at its two output ports in a versatile and reconfigurable manner. We identify a set of such functions and explain the reasons for their selection according to their role and significance. We define performance criteria and adapt them to the special nature of the logical outcome of each target function. We provide simulation results concerning the impact of critical operating parameters on these metrics and based on their concise interpretation we give guidelines for the execution of each Boolean function both with logical correctness and high quality. We also address several practical issues regarding the real implementation of these functions. Finally, we report on characteristic lightwave digital applications of enhanced functionality enabled through complex circuits, which are synthesized and formulated by combinations of the considered QD-SOA-based MZI Boolean functions.

INTRODUCTION

Modern telecommunication networks are characterized by the intense generation, use and storage of enormous amount of information. This trend has been rapidly established by Internet-related services and is further being fuelled by new developments, such as cloud-computing, data centric systems, video on demand, online gaming and intelligent mobile applications. In order to cope with the continuously increasing traffic that manifests in this broadband environment and satisfy the diverse users needs, it is necessary to create a platform of enhanced data-carrying capability. The main technological path followed for this purpose is to adopt optical fibers as the physical medium owing to their inherently large bandwidth. However the tremendous capacity they offer cannot be conveniently handled by the available electronic circuitry. This mismatch occurs due to fundamental obstacles that are encountered with the upgrade of single channel data rates into the sub-Tb/s region [1] and beyond [2]. This so-called 'electronic bottleneck' can be overcome by avoiding, where possible, the cumbersome optical-to-electrical-to-optical conversions and the concomitant increased cost, power consumption and complexity. The achievement of this goal relies on successfully performing several signal processing tasks entirely in the optical domain, i.e. all-optically, and hence directly at the ultra-high speed optical line rate [3]. Then, just as in electronics, the key requirement in this effort is the ability to execute Boolean functions so as to ensure the minimum degree of logical functionality between lightwave modulated signals. Thus this chapter deals with how this can be done by employing the novel devices of quantum-dot semiconductor optical amplifiers (QD-SOAs) and exploiting their potential as ultrafast nonlinear elements when placed in the two arms of a properly configured and driven Mach-Zehnder interferometer (MZI). In this context we present results obtained from the appropriate modeling of this switching module, which confirm the feasibility of implementing all-optically the AND, NAND, NOR and OR Boolean functions with logical correctness and high quality. We also address the critical issue of constructing digital logic architectures of advanced level by interconnecting multiple stages of such Boolean functions, and to this scope we report on an exemplary case that falls within the field of optical computing [4].

Representative State-of-the-Art Technologies for All-Optical Implementation of Boolean Functions

Figure 1 depicts the generic block diagram of a switching module that executes all-optical Boolean functions. The achievement of this goal requires employing a suitable medium, which must present a nonlinear behavior in response to the perturbation of its optical properties by the applied data signals. For this purpose it can be used not only as a discrete entity but also incorporated in an interferometric configuration. Depending on the target logic function and the exploited nonlinear phenomenon it may be necessary to utilize one or more additional signals. These extra inputs can be streams of repetitive pulses or can contain binary information, while they suffer the effects of the nonlinear phenomena that manifest inside the corresponding medium when the data signals propagate along the host nonlinear device. The nonlinear medium must be judiciously chosen since its inherent characteristics define decisively the quality of the all-optical Boolean operation.

Various technologies have been proposed in order to implement Boolean functions in an all-optical way [5]. Nevertheless, those being eligible for consideration as competitive must meet some indispensable criteria which include: a) the ability to support operation at ultrafast data rates, of the order of 160 Gb/s and beyond, so as to comply with modern broadband communication trends and needs, b) the ability to handle optical pulses of return-to-zero (RZ) data format, which is widely employed in real lightwave networks due to its superior transmission properties [6], c) the ability to perform logical inversion and subsequently generate the complement of a data signal involved in a Boolean operation. As it will be explained in the following of this chapter this conjugate binary signal can be used too as input to the switching configuration employed for this task [7], d) the adequate degree of maturity as proved by the repeated use in various sophisticated applications, and e) the potential to create a uniform platform having as core building module an all-optical gate whose role will not be limited to the realm of Boolean logic but will be extended to a broader set of fundamental and advanced all-optical signal processing functionalities. These conditions can be primarily satisfied by i) highly nonlinear fibers (HLNF), ii) periodically poled lithium-niobate (PPLN) waveguides and iii) conventional SOAs.

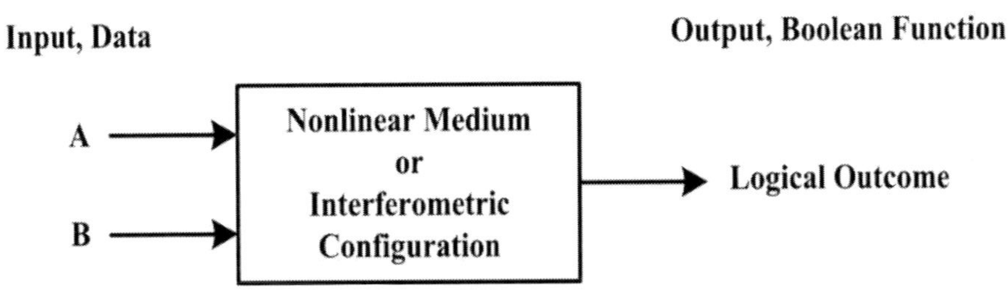

Figure 1. Generic block diagram for execution of all-optical Boolean functions.

More Specifically:

i) Highly nonlinear fibers (HLNF)

Highly nonlinear fibers (HLNF) allow to implement all-optically a Boolean function at an ultrafast data rate owing to the femtosecond response of Kerr effect in silica [8]-[9]. This is done by exploiting the nonlinear effect of cross-phase modulation (XPM), either with the help of optical bandpass filtering [8] or by placing the HNLF in a nonlinear optical loop mirror [9]. However, the simplicity and ultrafast potential of this solution is compromised by the fact that for efficient all-optical switching a quite high peak power is required and thus expensive and power consuming erbium-doped fiber amplifiers (EDFAs) must be inevitably used. Moreover, the length of the HLNF is such that it inhibits integration and increases latency. Therefore the adoption of HLNF-based schemes of all-optical Boolean functions is less convenient for the purposes of forming more complex circuits of enhanced combinational and sequential functionality, especially with feedback.

ii) Periodically poled lithium-niobate (PPLN) waveguides

These devices exhibit high nonlinear coefficient, ultrafast response time, negligible spontaneous emission noise, complete bit rate transparency and compactness, which make them promising candidates for executing all-optical Boolean functions [10]. The physical mechanism on which they rely is a combination of sum- and difference-frequency generation (SFG/DFG) when two pump signals interact in this medium. PPLN waveguides have proven capable of realizing Boolean functions in the optical domain up to 160 Gb/s [11]. However the PPLN-based logic schemes which operate by cascading SFG/DFG and SFG parametric processes while properly exploiting the phase matching characteristic requires to use many light sources [12]. This in turn demands supplying excessive pump power and accordingly operating at unrealistically high temperatures to avoid photorefractive damage, which means that more power must be consumed to heat the device [13]. Also they suffer from low conversion efficiency as a trade-off for increasing the limited bandwidth within which SFG only occurs [13]. Furthermore resorting to pump depletion phenomena in order to increase bandwidth [10] presents additional complexity since it requires to synchronize a strong clock pump with the input signal [13].

iii) Semiconductor optical amplifiers (SOAs)

Among the technological approaches that have been devised for the all-optical implementation of Boolean functions those that rely on SOAs are very popular. This particularly holds when these active devices are encompassed by an interferometric configuration, since then a set of attractive features is offered [14]. These include signal amplification, wide gain bandwidth, low latency, reasonable energy consumption, multiple terminal operation, regeneration capability, noise and jitter tolerance, stability to environmental changes and the potential for photonic integration with the subsequent advantages of compact construction, enhanced reliability and affordable cost.

The operation of these SOA-based interferometric schemes exploits the well efficient nonlinear phenomenon of cross-phase modulation which manifests in a SOA under the influence of one or two strong data modulated optical beams acting as control signals [15]. Depending on whether the pulses of these signals are both present or absent or are available by turns, a phase difference is induced between the copies of another, weaker optical signal, which if it is also pulsed then it plays the role of the clock. This process is translated into a change in the intensity of the signal that is switched and subsequently into the achievement of a specific Boolean function either between these control signals or between the existing control signal and the clock signal.

THE NEED FOR QUANTUM-DOT SEMICONDUCTOR OPTICAL AMPLIFIERS

Despite the important advantages of SOA-based interferometers, the extension of their use to implement all-optical Boolean functions at ultrafast data rates is limited by the inherently slow SOA gain recovery time. The negative by-product of this fact is that it is not possible to satisfy the unceasing demand for more bandwidth since single channel data rates are upgrading to magnitudes that exceed by far the SOAs ultrafast potential [1] and this trend is expected to continue in the future. The reason for this impediment is that when the SOA is driven by a strong data stream comprising of discrete pulses whose repetition interval is smaller than the SOA carrier recombination rate, as it happens in most practical cases, then the change of its gain dynamics induced by this signal (control) is not uniform from pulse to pulse but depends on the binary content of the inserted information. The negative by-product of this irregular gain variation is the appearance of the so-called pattern effect [16], which distorts the creation of the proper differential phase shift between the pairs of components of the pulses that belong to the signal to be switched (clock). Consequently this impairment degrades the quality of the logical outcome of the SOA-based interferometer, which manifests as amplitude fluctuations of the marks and incomplete suppression of the spaces.

In order to confront this problem various ideas have been proposed, which aim, each in a different way, at accomplishing the same thing, namely to accelerate the process of SOA gain recovery so that its dynamic behavior and subsequently its response depends solely on the current input bit and not on the preceding ones. More specifically, the most direct means consists in intervening in the operation and design of the SOA through the increase of the injection current [17] and the change of the dimensions of the active region [18], respectively. In the first case it is necessary to use power consuming and costly current sources while altering the bias current may distort the shape of the pulses that transit the SOA [19]. The second case concerns either the increase of the length [17] or the optimization of the waveguide cross section [18], which are accompanied by consequences that cannot be ignored. These include the fact that beyond a permissible length the amplified spontaneous emission becomes significant [17] and adds as noise to the amplified signal, thereby compromising its quality. Also depending on the topology of the interferometric arrangement and the relative direction of propagation of the control and clock signals it may not be feasible to create the necessary phase shift and switching window [20]-[21]. Furthermore, the minimization of the optical confinement area affects the SOA saturation energy [18], which however determines the amount of power that must be provided to the SOA so as to achieve correct and good switching. Concurrently this action can increase the induced phase shift [22] much above the desired degree, thus causing the distortion of the switched pulses [23]. This essentially means that from the SOA perspective the satisfaction of the set performance criteria is not guaranteed because due to the correlation of the involved parameters the change of one of them can disturb the other, and vice versa. An alternative technique is to simultaneously launch a continuous wave (CW) strong holding beam [24] or an assist light near the SOA transparency point [25], which reduces dramatically the SOA carrier lifetime from typically several hundred to a few picoseconds [26], while the transfer function of the interferometer is converted from cosinusoidal to a hard-limiting form [27], which is beneficial for smoothing the amplitude fluctuations to the same level. Despite its efficiency, in practice the implementation of this technique is not a straightforward or easy task since, besides the price of providing a relatively high power from laser sources, special care must be taken concerning the adjustment of the power and the selection of the wavelength of the CW signal so as to compensate the inevitable small signal gain reduction and still be possible to create the desired phase shift [28]-[29]. Another approach does not deal with the problem at its root but indirectly based on different filtering methods to enhance the speed of the SOA response. The common characteristic of these methods is that they attempt to balance the dynamic changes of the SOA gain and phase through the selective passage or elimination of spectrally broadened portions of the signal to be switched [30]. In this category fall for example the optical delay interferometer [31]-[32] and the so-called 'Turbo' switch [33]-[34], which although they contribute significantly, yet they do not lack limitations. Thus in the first type of filter the removal of part of useful information due to the filter detuning against the carrier frequency inserts losses and inevitably reduces the signal-to-noise ratio, whilst the choice of the delay for alleviating the pattern effect is made more difficult the faster the data arrive since the maximum value of this temporal parameter must not exceed half the data period [35]. In that case it is necessary to use or combine more sophisticated filters [36]-[37] whose transfer function should be designed in terms of its shape and slope in such way that a good trade-off is achieved with respect to the satisfaction of diverse performance criteria, which is not an easy task at all. On the other hand the 'Turbo' switch uses one or more additional SOAs which are connected in series to the initial one with passive filters placed between them and which, by acting as nonlinear filters, compensate for the slow gain recovery component, thus accelerating the response of the whole combination. However, and despite the remarkable potential of this method, its main disadvantage is the lack of flexibility when it is applied for all-optical switching purposes since the existence of extra SOA stages, except for causing undesirable noise accumulation, renders difficult the process of practical handling and implementation. This fact has a negative impact on the unobstructed exploitation of the method in more complex all-optical circuits and subsystems which are based on optical switches and whose operation may be inherently demanding, as for example due to the existence of feedback paths [38]. Finally, a widespread mechanism is based on a differential scheme, where the signals that control switching enter the gate with the appropriate temporal succession and interleaving so that the width of the switching window is determined by the deliberately imposed relative delay, whose value is chosen to be smaller that the SOA carrier lifetime [39]-[41]. Nevertheless this way is rather tricky since the achievement of accurate synchronization is not an easy task while in the most extreme case each of the SOAs receives two pulses in short succession in a single bit slot, which imposes a heavy strain on their gain dynamics [39] and actually increases their susceptibility to the pattern effect.

From all the above arguments it is realized that the solutions adopted in the effort to mitigate the SOA pattern effect present serious technical constrains, which do not allow SOA-based interferometric switches to fully serve the purpose for which they are destined. Inevitably these modules cannot keep efficiently pace with the excessively increasing line rates, as they can support only a small fraction of those that are currently underway [2] and soon or later will be the norm in future optical networks. It is thus clear that there is a need to follow a different and novel technological path, which should have the perspectives of successfully responding to this modern challenge in a more direct and affordable manner. In this context quantum-dot (QD) SOAs with their special structure are very promising devices thanks to their distinctive physical properties that have been constantly improving during recent years over conventional SOAs as a result of the intense research carried-out in this field. These include the lower current density threshold, the higher saturation output power, the wider gain bandwidth, the lower noise figure, the similarly low polarization gain dependence and the weaker thermal dependence [42]-[44]. However their most important characteristic is their exceptional faster gain recovery time [45], which enables them to respond much quicker to ultrafast data compared to SOAs that don't have QDs inserted in their active region. These combined advantages have made QD-SOAs ideal candidates for exploitation in all-optical signal processing applications with high performance against the pattern effect as well as other metrics [46]-[47]. One such important application concerns the accomplishment of all-optical Boolean functions by incorporating QD-SOAs in interferometric architectures, and among the principal existing options [48] the Mach-Zehnder interferometer (MZI) is most suitable as it effectively combines the structural simplicity, the co-propagation of the launched signals, the possibility to independently control the phase in each arm, the compact size, the thermal stability, the low switching energy requirement, the regeneration capability, with the potential of integration and the overall practicality [49]. Owing to these attractive features the QD-SOA-based MZI has recently been the primary choice for the demonstration of all-optical logic functions [50]-[53].

QD-SOA STRUCTURE

The schematic diagram of a QD-SOA is shown in Figure 2(a). This device is equivalent to a conventional SOA in terms of its basic structure and applications but its active region consists of multiple QD layers [16], [54]. The fabrication of these QDs is based on a self-assembled growth technique called Stranski-Krastanov. Initially this involves the growth of a two-dimensional layer, namely the wetting layer (WL). When a critical layer thickness is exceeded, then three-dimensional islands are formed on top of this layer [55]-[56]. QD-SOAs operating in the 1550-nm range consist of InAs QDs grown on InP substrate [45], [57]. Assuming that the conduction band of a single QD has two discrete electronic states, namely the ground state (GS) and the exited state (ES), and that the wetting layer offers an additional energy level (WL), a QD-SOA is considered to be a three level system. Owing to the discrete nature of the energy band diagram of the QDs, QD-SOAs acquire special dynamic characteristics and exhibit remarkable optical properties. Figure 2(b) depicts the energy band diagram of the QD system and the WL, as well as the basic electron transitions between the three energy states of the conduction band. Electrons are assumed to limit the carrier dynamics, since holes have larger effective mass resulting in smaller level spacing and thus faster relaxation [42], [58]. The carrier capture directly into the GS is not taken into account due to fast intradot carrier relaxation and large energy separation between the GS and the WL band edge [42]. Furthermore, the holes are assumed to be in quasi thermal equilibrium at all times [42], [58]. Therefore the full set of transitions includes [52] (a) the electron transitions between the WL and the ES, which are characterized by the transition times τ_{w2} and τ_{2w}, (b) the electron transitions between the ES and the GS, which are characterized by the transition times τ_{21} and τ_{12}, and (c) the spontaneous recombination in the WL and the QDs. In particular carriers are injected into the WL through electrical pumping and due to phonon- and Auger-assisted processes [16], [42] they relax to the ES. Similarly carriers from the ES relax to the GS, while the amplification process is incited by the stimulated radiative transition from the GS of the conduction band to the GS of the valence band. In this manner the states above the GS serve as a carrier reservoir for the GS [16], [57]-[58], which accelerates the QD-SOA gain recovery and accordingly its response to ultrafast excitations.

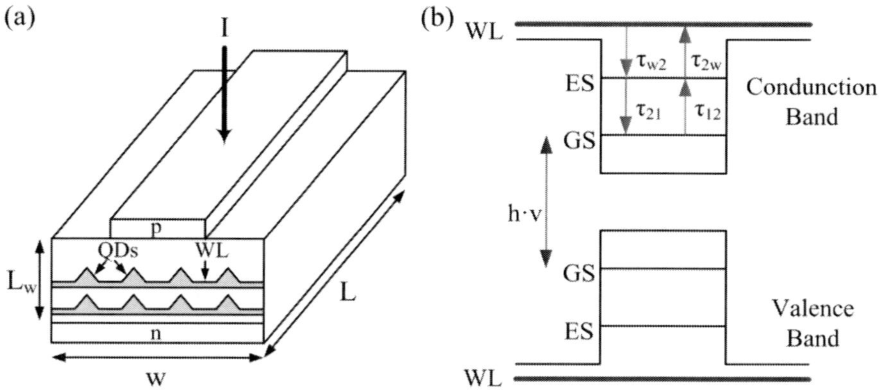

Figure 2. (a) Schematic diagram of a QD-SOA, (b) Energy band diagram of the QD system and the WL.

QD-SOA Modeling

Depending on the target Boolean function, the driving conditions of the involved QD-SOAs may differ concerning the type and number of input signals. Thus the QD-SOAs modeling must be adaptable to this fact so as to theoretically describe the particularities of each case in an appropriate way. For explanatory reasons we consider the most generic but also simple situation where a QD-SOA receives only one data signal whose power is capable of modifying the QD-SOA optical properties. The employed model takes into account the propagation of this pulsed signal along the longitudinal direction, z, of the QD-SOA in conjunction with the change of the QD-SOA gain dynamics. In this context we treat the QD-SOA as a three-level system according to the energy band diagram described in the previous section. This approach allows us to obtain realistic, accurate and global results. It also complies with the relevant finding that the major contribution to the ultrafast gain recovery of QD-SOAs is made by the cascade of the dynamical process, which includes indirect relaxation from the WL via the ES to the GS [59]. In comparison, if it were assumed that the QDs' conduction band, and accordingly its valence band, consists of the GS only, then although the influence of the direct relaxation channel could be explicitly taken into account, yet the carrier kinetics could not account quantitatively for the exceptionally fast QD-SOA gain recovery [59]. However it is exactly owing to this unique property that QD-SOAs can be exploited as nonlinear elements for executing all-optical logic functions at sub-Tb/s data rates. Moreover, the absence of the intermediate ES state is replaced by calculating the carrier occupation probability near the band edge of the WL, which in turn is populated by the injection current and serves as a reservoir of carriers. As a consequence this modeling approach is valid when the population of carriers in the WL is large [60]. Hence it is restricted to specific QD-SOAs driving conditions and leads to more favorable predictions for the capability of QD-SOAs to be incorporated in interferometric configurations for all-optical logic purposes [50]. Therefore the 3-level rate equation model (3LREM) that we have adopted for the QD-SOAs dynamical behavior is mostly suitable for simulating the operation of the QD-SOA-based MZI and subsequently of the Boolean functions which are implemented using this switching module.

Following [61], the model is formulated by the propagation equation (1) and the 3-level rate equations (2)-(4) for the wetting layer (WL), excited state (ES) and ground state (GS), respectively:

$$\frac{\partial S}{\partial z} = \left[g_{max} \left(2f - 1 \right) - \alpha_{int} \right] S \tag{1}$$

$$\frac{\partial N_w}{\partial t} = \frac{J}{eL_w} - \frac{N_w(1-h)}{\tau_{w2}} + \frac{N_Q h}{L_w \tau_{2w}} - \frac{N_w}{\tau_{wR}} \tag{2}$$

$$\frac{\partial h}{\partial t} = \frac{L_w N_w \left(1-h\right)}{N_Q \tau_{w2}} - \frac{h}{\tau_{2w}} - \frac{\left(1-f\right)h}{\tau_{21}} + \frac{f\left(1-h\right)}{\tau_{12}} \tag{3}$$

$$\frac{\partial f}{\partial t} = \frac{\left(1-f\right)h}{\tau_{21}} - \frac{f(1-h)}{\tau_{12}} - \frac{f^2}{\tau_{1R}} - \frac{g_{max}\left(2f-1\right)L_w V_g}{N_Q} S \tag{4}$$

These equations hold for a negligible QD-SOA facet reflectivity and amplified spontaneous emission (ASE) [51], [53], which, as explained and justified in detail in the section of the chapter that deals with these practical issues, are valid assumptions that greatly simplify the model's formulation. In relationships (1)-(4), variable t is time transformed into a retarded frame moving with the pulse [52] and the functions involved in the derivatives are the photon density of the driving data signal, $S = S(z,t) = P(z,t)/(A_{eff}V_g h\nu)$ (where P is power, $A_{eff} = 0.75\ \mu m^2$ is the effective cross section of the QD-SOA, $V_g \approx 8.3\times10^7 m/s$ is the group velocity of the propagating signal and $h\nu$ is the photon energy), the electron density in the WL, N_w, and the electron occupation probabilities in the ES and GS, h and f, respectively. Also $g_{max} = l\frac{2N_Q}{L_w}\sqrt{\pi \ln 2}\frac{\gamma_{hom}}{\gamma_{inhom}}\sigma_{res}$ is the maximum modal gain, where l, N_Q and L_w are the number, surface density and effective thickness of the active QD layers, respectively, γ_{hom} is the homogeneous linewidth of the resonant QDs, γ_{inhom} is the inhomogeneous linewidth of the QD ensemble and σ_{res} is the resonant cross section of the carrier-photon interaction [62]. Furthermore, α_{int} is the material absorption coefficient, J (= I/(wL), where I is the applied current, w is the width of the active layer and L is the QD-SOA length), is the injection current density, e is the electron charge, τ_{w2} is the electron relaxation time from the WL to the ES, τ_{2w} is the electron escape time from the ES to the WL, τ_{wR} is the spontaneous radiative lifetime in the WL, τ_{21} is the electron relaxation time from the ES to the GS, τ_{12} is the electron escape time from the GS to the ES and τ_{1R} is the spontaneous radiative lifetime in the QDs.

The system of coupled differential equations (1)-(4) is numerically solved in a step-wise manner for pulses that belong to a pseudo-random binary sequence (PRBS) frame of the data signal. For this purpose each pulse is sampled over its period, T_{per}, at discrete temporal intervals, $\Delta t = T_{per}/k$, and the QD-SOA is divided into n spatial segments of uniform length, $\Delta z = L/n$, as shown in Figure 3. In this manner a two-dimensional spatio-temporal grid, $\Delta t \times \Delta z$, is formed. The selection of k and n must be carefully made so that they are integer numbers and at the same time they ensure a high temporal resolution and computational efficiency. A trade-off can always be made between the two latter requirements depending on the available memory and the processing speed of the computer machine. In particular the number of longitudinal sections must be larger than 10. Otherwise the amount of the differential gain and phase shift induced in the QD-SOA by the signal that controls switching is overestimated and underestimated, respectively, thereby impeding the correct modeling of the SOA-based interferometric configuration [63]. This critical condition affects in turn the choice of k, given that the infinitesimal space and time bins are related to each other by $\Delta z = V_g \Delta t$. Thus based on this relationship the number of spatial intervals is chosen $n \geq 10$, so that for a given Δt each temporal sample travels the same distance, Δz, inside the amplifier. Then by following a trial and error procedure for different data rates, it can be confirmed that the QD-SOA dynamics are properly simulated provided that $\Delta t \leq 106.25$ fs, which corresponds to a spatio-temporal grid smaller than 9.4×10^{-4} ps · mm.

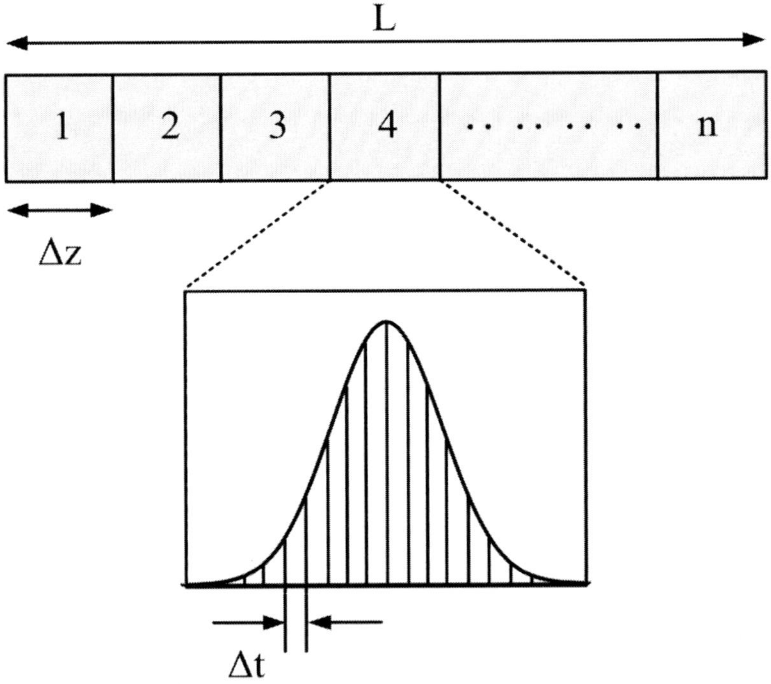

Figure 3. Spatio-temporal grid on which the 4th order Runge-Kutta method is applied. L: QD-SOA length. n: Number of uniform segments that QD-SOA is divided into. Δz : Length of each spatial segment. Δt : Temporal interval between two consecutive samples of an input pulse.

Taking all the above points into account, $\Delta t \times \Delta z = 3.125 \times 10^{-4} (5 \times 10^{-5})$ ps · mm is chosen for a 160 (320) Gb/s data signal, which means that $k = 100$ (125) and $n = 800$ (2000) for a typical QD-SOA length $L = 4$ mm. Then the 4th order Runge-Kutta numerical method is applied to find the QD-SOA amplification factor given by $G(t) = P(L,t)/P(0,t)$. For this purpose it is necessary to define the appropriate initial conditions for the four functions in (1)-(4). More specifically, for S this done through the equivalent power, whose profile is assumed to be Gaussian, i.e. $P(0, t) = P_{peak} \exp\left[-4\ln 2\left(t/T_{FWHM}\right)^2\right]$, where P_{peak} is the pulses peak power and T_{FWHM} is their full-width at half-maximum, which corresponds to some fraction of the operating period. For h and f, on the other hand, this is done by observing that before any signal is launched, there is no change in the QD-SOA gain dynamics and so their initial values equal unity. Concurrently the QD system is at equilibrium and so the temporal derivatives in (2)-(4) are nullified. Then by replacing $h = 1$ in (2), the initial value of N_w can be found for known values of the other parameters.

The required programming algorithm was developed in Matlab code. With the aforementioned choice of k and n, the execution of the model at 160 Gb/s is completed within approximately 70 seconds. This running time is achieved in a desktop computer with an Intel® Core™ i5-650 3.20 GHz processor, 3.5/4 GB use of RAM (DDR3 PC3-10700 667MHz) and Microsoft Windows 7 Home Premium SP1 32-bit operating system. Unless otherwise mentioned, and excluding the case where a QD-SOA parameter is varied in order to assess its impact on performance metrics, the numerical simulation is carried out for typical QD-SOAs parameters values taken from the available literature on various interferometric gates based on QD-SOAs [51]-[53].

Last but not least, modeling features addressing the simultaneous propagation of multiple signals along the QD-SOA need to be discussed. As already mentioned, the all-optical implementation of a target logic function might entail launching additional signals that suffer the nonlinear influence of the data signals. For the proposed QD-SOA-MZI-based solutions this refers to extra signals of properly weak intensity. Since these signals do not provoke changes to the QD-SOA dynamics, this implies that they do not affect directly the system of coupled differential equations (1)-(4). However, their ability to undergo the data-induced nonlinear behavior depends on their spectral offset from the data signals. For this reason the homogeneous broadening is physically taken into account, as it determines the maximum allowed detuning of the involved signals. A special consideration of this effect would be necessary if the pursued goal were to perform some multi-wavelength task [64], where different input channels that are located within the homogeneous broadening interact strongly via cross-gain modulation (XGM). This nonlinear effect can also manifest if the detuning between the data and additional signals is sufficiently large compared to the inhomogeneous broadening, which occurs due to fluctuations in QD size and material composition [46]. In this case the QD ensemble should be divided into different groups with energies depending on their resonant frequencies. Then the QD-SOA dynamical behavior should be described by properly modifying the system of 3-level rate equations [58] while discriminating the modal gains of the different co-travelling signals by assuming a Gaussian-like distribution for their spectral profiles [57]. Nevertheless, although this possibility of XGM in QD-SOAs would be beneficial for multi-wavelength applications [46], yet the switching scheme we have proposed aims at realizing Boolean logic operations on single channel ultrafast data rates and hence at supporting all-optical signal processing in the context of Optical Time Division Multiplexing (OTDM) [1].

Classical QD-SOA-Based MZI

Figure 4 depicts the classical architecture of the QD-SOA-based MZI.

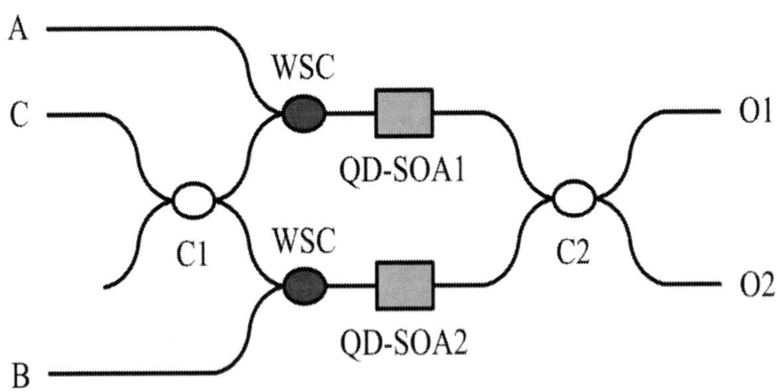

Figure 4. Classical architecture of QD-SOA-based MZI.

It consists of a MZI with QD-SOAs placed symmetrically in its arms, which are formed by joining together two 3 dB couplers, C1 and C2. Two data streams A and B are inserted in the upper and lower MZI arms through the homonymous input ports, respectively, with the help of a wavelength selective coupler (WSC). Concurrently a clock stream (CLK) held continuously to a logical '1' enters the structure via input port C and is split by input 3 dB coupler, C1, into a pair of equal parts, which travel distinctly along the identical QD-SOAs located in their path. In order to be able to achieve switching, A and B should be at least an order of magnitude stronger than the CLK while their wavelength detuning from the CLK in the 1550 nm band, as described in the previous section, should lie within the homogeneous broadening of the two QD-SOAs [51]-[53]. In this manner signals A and B can properly modify the nonlinear optical properties of the respective QD-SOAs and induce through XPM a phase difference between the decomposed CLK components. Furthermore, since data signals A and B act independently on different nonlinear media, it is possible to be positioned at the same wavelength, thereby decreasing the number of laser sources which are required for providing them.

Now if A and B carry the same binary information then when A = B = '0' the CLK replicas are unaffected and when A = B = '1' they suffer the same QD-SOA dynamic influence. Thus in both cases no relative phase difference is created between them and on their combination at the exit 3 dB coupler, C2, no output is produced, which is equivalent to a logical '0' at port O1, or direct switching port, of the MZI. In contrast if the logical states of A and B are different, i.e. A = '1', B = '0' and vice versa, then the divided CLK constituents acquire through XPM a differential phase shift and the initial phase balance of the MZI is broken. If this phase difference is made equal to π through suitable design of the MZI then the CLK copies interfere constructively at output O1 and a logical '1' occurs therein. In each case the outcome of the Boolean operation between the four logical combinations of A and B is imprinted on the CLK and is transferred to the output O1, where a bandpass filter (not shown) allows only the switched signal to pass through and rejects A and B.

Modified QD-SOA-Based MZI

The devised circuit that we have used for the all-optical design of Boolean functions combines the MZI architecture, which is modified so that the MZI is transformed into a Fredkin gate [65]. In its generic form, which is shown in Figure 5 in expanded setup (a) and block diagram (b) form, it comprises of four input ports, A, B, C and D, and two output ports, O1 and O2. Among them, inputs C and D are used to insert in the interferometric arrangement two binary signals which are incoherent to each other, so that they do not interfere [66]. These signals are split via the 3 dB input coupler C1 into a pair of equal parts which travel distinctly along the identical QD-SOAs located in their path. At the same time inputs A and B are connected to wavelength selective couplers (WSC) and are used in order to launch into the upper and lower MZI arm, respectively, the data-carrying signals on which the intended Boolean functions are applied. Provided the magnitude and wavelength detuning conditions described in the previous section hold against C and D, these signals become capable of modifying the optical properties of the respective QD-SOAs and inducing through cross-phase modulation (XPM) a phase shift on the divided copies of both C and D. In this manner when the latter recombine and interfere at the 3 dB coupler C2, the relative phase difference that they have acquired through this nonlinear effect is converted by the MZI into an amplitude change at the output, which in turn is translated into binary information. According to this principle of operation the exact logical product at output port O1 depends a) on the combination of the binary content of A and B which are responsible for switching and b) on the existence of signals C and D which undergo the change of the QD-SOAs gain dynamics by A and B and on which the result of this action is imprinted. In the first case, a), the phase balance between the MZI arms is broken only if either A or B is ‘1’, otherwise if both are ‘0’ or ‘1’ it is preserved. The physical reason for this behavior is that when A and B differ they exert a differential influence on the optical phase condition of the MZI, whilst when they are identical their contribution is null or equal. However this dual rail ‘‘any but not all” effect is not sufficient on its own to alter the state of the MZI, unless it is enabled by C and D. Then, and referring to the second case above, b), the final form of the logical outcome at output O1 is determined by the way this port is interconnected to the ports from which signals C and D enter the setup individually. Thus since O1 is the direct switching port for a signal coming from input C and the cross switching port for a signal coming from input D, this essentially means that due to the conservation of energy the respective outcome obtained therein is the logical complement of each other. As a consequence when C and D are superimposed the total output will be logically described by the sum [65]

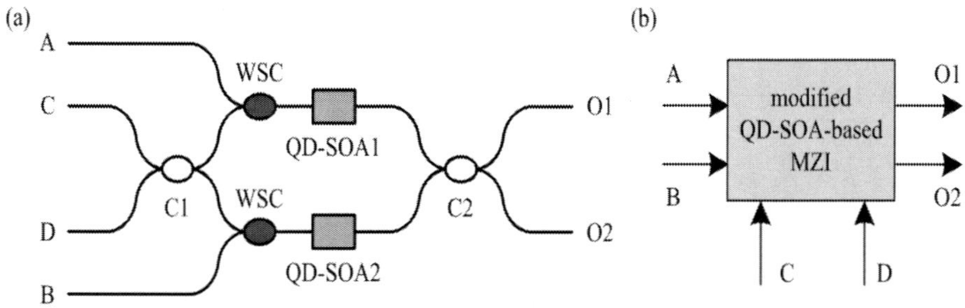

Figure 5. Schematic (a) and block (b) diagram of modified QD-SOA-based MZI.

$$O1 = C(A \oplus B) + D(A \odot B) \tag{5}$$

where the symbol $\oplus$ denotes modulo-2 addition and $\odot$ the conjugate. *This* expression suggests that if C and D are suitably chosen and combined then it is possible to obtain at output O1 the truth table of the desired Boolean function between A and B.

Modified QD-SOA-Based MZI Modeling

According to (5), the operation of the QD-SOA-based modified MZI is theoretically described from

$$P_{O1}(t) = T_{O1,C}(t)P_C(t) + T_{O1,D}(t)P_D(t) \tag{6}$$

where P_{O1} is the power of the signal that emerges at output port O1 and which carries the result of the executed Boolean function, while P_C and P_D are the powers of the signals inserted into the MZI from ports C and D, respectively. Furthermore, $T_{O1,C}(t)$ and $T_{O1,D}(t)$ are the transfer functions of the MZI in response to inputs C and D, which are given by

$$T_{O1,C}(t) = \frac{1}{4}\left\{G_1(t) + G_2(t) - 2\sqrt{G_1(t)G_2(t)} \times \cos\left[-\frac{\alpha_{LEF}}{2}\ln\left(\frac{G_1(t)}{G_2(t)}\right)\right]\right\} \tag{7}$$

$$T_{O1,D}(t) = \frac{1}{4}\left\{G_1(t) + G_2(t) + 2\sqrt{G_1(t)G_2(t)}\times\cos\left[-\frac{\alpha_{LEF}}{2}\ln\left(\frac{G_1(t)}{G_2(t)}\right)\right]\right\} \quad (8)$$

where α_{LEF} is the QD-SOAs linewidth enhancement factor, which determines the level of phase shift that can be achieved through XPM inside each QD-SOA and which according to findings from investigations on real QD-SOAs operating in the 1550 nm wavelength window it can take high values [67]. Then from (6)-(8) we observe that in order to calculate the light intensity at output port O1 we have to replace the values of the time-dependent gains experienced by the copies of the weak signals in QD-SOAs 1 and 2, $G_1(t)$ and $G_2(t)$, respectively. These values can been found according to the previously described numerical procedure, which is followed for solving the system of equations (1)-(4) for each QD-SOA and for the input signals involved in the implementation of the target Boolean function.

Real Implementation Issues

The actual all-optical implementation of the reported Boolean functions requires for completeness to take into account some practical issues and how these may affect their design and applicability according to the guidelines derived from the theoretical analysis of their operation. These issues include the QD-SOAs amplified spontaneous emission (ASE), polarization gain dependence (PDG), reflectivity of the front and rear facets and fiber coupling losses, the coupling ratios of the 3 dB couplers and WSCs, the power consumption and the availability of the logical complement of the launched data.

First, concerning real QD-SOAs, an undesirable effect that may compromise their operation is that of amplified spontaneous emission (ASE). This effect may be detrimental for their performance and consequently for the switching module in which they are incorporated for executing a Boolean function. This happens because it manifests as noise which is added to the output signal and degrades its quality [68]. The exact evaluation of this negative contribution of ASE is a rather demanding modeling task [69] due to its complicated nature. In fact most relevant treatments have been conducted so far on linear QD-SOA amplification as well as on QD-SOA static and not dynamical behavior [70], which however are not applicable to the case of all-optical interferometric switches with QD-SOAs. Nevertheless the impact of ASE on the performance of the QD-SOA-based MZI when configured and driven for executing Boolean functions can be qualitatively estimated based on some of the information available on SOA ASE. More specifically, it is known that the amount of differential gain modulation that must be induced for a weak signal to be switched by the action of another stronger signal in the nonlinear medium of a SOA with ASE deviates from the ideal situation without ASE [69]. The consequence of this discrepancy is that the excursions of the peak amplitudes of the marks cannot be so efficiently made to occur around the same average level, which increases the amplitude modulation. Similarly in the presence of ASE the phase difference for obtaining full destructive interference is not minimized due to the existence of ASE noise phase variations in both QD-SOAs that change randomly [71]. The negative by-product of this impairment is that the spaces cannot be completely eliminated, which decreases the extinction ratio. Therefore it would seem expectable that subject to ASE the operating requirements for ensuring equally acceptable performance of the Boolean functions as when this phenomenon is ignored would become comparatively tighter. However this is not allowed to happen because the inversion or spontaneous emission factor, which determines the contribution of ASE to the switched signal [72], can be made in QD-SOAs to approach unity and hence the corresponding ASE noise level can be kept minimum. This is possible owing to the existence of the carrier reservoir in the QD-SOA structure, and the enhancement of the degree of population inversion in the active electronic QD states for moderate currents injected to medium QD-SOA lengths [42], [73]. Furthermore, according to the experimental characterization of QD-SOAs gain dynamics around 1550 nm the contribution of ASE is strongly reduced when the data pulse repetition interval is comparable to the recovery time of the discrete ES level contained in the QDs [74]. Both simulation [75] and pump-probe measurement [76] results on the evolution of the QD-SOA carrier dynamics during amplification of a strong input pulse have revealed that the ES level recovers on a time scale of several picoseconds. This means that the aforementioned temporal condition is satisfied for Boolean functions executed at 160 Gb/s and above, as in the Boolean functions considered in this chapter.

But even in the worst scenario that ASE noise would be added to the power content of the data marks and spaces, their mean level that would result from their peak fluctuation would occur at the slope of the MZI sinusoidal transfer function, where they would perceive maximum and null transmittance, respectively [77]. In this manner the action of ASE would be compensated for and the noise contained in these bits would be suppressed. This in turn would not allow them to become distorted, which would be reflected on the quality of the outcome of the Boolean functions and of the performance metrics. Finally, narrow bandwidth optical filters could be placed after each QD-SOA in the MZI arms to reject further the out-of-band ASE noise, also known as spontaneous-spontaneous beat noise [78].

Thus based on all this reasoning, it can be realized that ASE does not impair the process of selecting the parameters which are critical for executing the all-optical Boolean functions using the QD-SOA-based MZI. This justifies ignoring ASE in the formulation of the model used to simulate the operation of these functions, which enables to greatly reduce the computational complexity while preserving the validity of the obtained results.

Second, the polarization dependence of the gain (PDG) of a QD-SOA seemed until a couple of years ago to pose a serious restriction in the effort to make feasible their exploitation in real applications. For example in all-optical logic function schemes the (absolute) difference between the gains along the transverse electric (TE), and transverse magnetic (TM) axes that results from this effect [79] makes the TE mode of light from a weak clock signal, which is injected into a SOA excited by a strong optical control signal, to suffer a change in its gain dynamics different from that of the TM mode [80]. As a consequence these orthogonally polarized components acquire a phase shift, which is a function of the TE and TM gains and builds up with propagation through the (QD-) SOA to an extent that it may not be negligible [81]. In this case the output power of the polarization sensitive QD-SOA will also depend on the phase difference induced due to PDG and not only due to XPM, as it should for proper switching. This however may disturb the operation and degrade the performance of the interferometric configuration in which the QD-SOA is employed. In QD-SOAs this problem would occur due to the growth of the QDs by the conventional Stranski-Krastanov method, which makes the optical properties of these devices to inherently exhibit a strong polarization gain dependence (see relevant references in [82]). Nevertheless this drawback has been now overcome by recent developments and the optimization of the QD shape and strain using the so-called strain-controlled columnar QD structures (SC-CQDs) [82]. In this manner it has been feasible to obtain a PDG as low as 0.4 dB and subsequently ultrafast polarization-insensitive error-free operation in the 1550-nm wavelength band while preserving the other QD-SOA attractive features. This significantly reduced value of PDG in QD-SOAs has enhanced their degree of polarization insensitivity and accordingly that of the interferometric switch, like the MZI, in which they are incorporated as nonlinear elements. This means that the process of switching is not impaired by this effect, and so the regularities revealed for the critical parameters of the Boolean functions should reliably hold when these are implemented in reality. As a consequence the approach to model the QD-SOAs as polarization independent devices is also justified and helps deliver a reliable outcome in a computationally accurate and efficient manner.

Third, we address a couple of QD-SOAs fabrication technicalities, namely the reflectivity of the front and rear facets and the fiber coupling losses. In more detail, the existence of optical feedback within the cavity of (QD-) SOAs may have a negative impact on their gain, saturation and noise characteristics as well as on their dynamic response [83] and the performance of nonlinear effects exploited for switching purposes [84]. Moreover, in order to account the influences and limits arising from this residual effect it is necessary to apply the relevant boundary conditions for the signal fields propagating in the forward and backward directions, which however requires a rather complex modeling. Fortunately the technology of QD-SOAs has evolved to the point that the reflection coefficients in real devices have been dropped well below 1 x 10^{-5} owing to the optimum combination of anti-reflection coatings and tilted waveguides [42]. The satisfaction of this demand is possible and sufficient for any QD-SOA length that provides a normal gain even under a high inversion of its states, which prevents them from start lasing. Thus it is permissible to leave facet reflectivities out from the simulation or equivalently assume that they are ideal and hence zero without compromising the realistic nature of our results. On the other hand the fiber coupling losses in (QD-) SOAs have been reduced down to 1 dB [43]. For this reason the values of the small signal gain used in the simulations of the all-optical Boolean functions are comparable to those of the net gain that is left after this kind of loss is subtracted from the maximum fiber gain of the device, and so the conditions specified for their appropriate operation are representative of the real world.

Fourth, we consider the effect of real passive elements used for inserting and exiting the signals involved in the Boolean operations, which include the 3 dB couplers for the clock and the wavelength selective couplers (WSC) for the control (data) signals. Notably in these elements energy is lost due to leakage instead of being converted into heat [85]. However if they are not ideal then they can impair several switching characteristics [86]. Nevertheless, due to the fact that there are two such couplers at the input and output of the MZI, the possible deviations from the 50/50 splitting ratio compensate each other to a sufficient extent [78]. Moreover, the commercial availability of 3 dB couplers with enhanced coupling ratio tolerance contributes decisively in increasing the robustness against uneven coupling and accordingly in preventing it from having negative implications. The situation is qualitatively similar for the data signals that control switching since the necessary WSCs are employed too in pair and any discrepancies related to coupling can be mutually neutralized. Furthermore, given that what actually matters is the power going into the respective QD-SOAs, the fine adjustment of the power supplied by the laser sources that generate these signals can also help preserve the level of saturation specified for high performance of each Boolean function. Therefore we can feel again confident that the simulation results reproduce well the requirements for the real implementation of these functions without being affected by these coupling imperfections.

Fifth, the power consumption of the QD-SOA-based MZI circuit used to implement the Boolean functions is very critical, since it is a measure of how efficiently it can make the most of the power offered to it both in optical and electrical form to achieve logic operation. This consumption can be quantified if we subtract the power that emerges at port O1, P_{O1}, from the sum of the powers that control switching, P_A and P_B, with twice the power supplied to each QD-SOA, $P_s = V_b I_b$, where V_b and I_b are the applied bias voltage and current, respectively. Then the total power consumption reads [87]

$$\text{POWER CONSUMPTION} = P_A + P_B + 2P_s - P_{O1} \tag{9}$$

This power corresponds to a switching energy per bit whose level lies in the range of typical values, since both the optical and electrical power consumption, which are of the oder of 100 fJ per bit the former and less than 10 pJ per bit the latter, agree with the trend reported for state-of-the-art all-optical digital logic modules that rely on SOAs as nonlinear elements at data rates as ultrafast as in our case [88]. However, from this expression we also note that the term that includes the injection current is dominant since, for a value of the order of 300 mA that corresponds to the mean current density specified from the simulation results and a typical applied voltage of the order of a few volts [88], its magnitude is well over that of signals A, B and O1, whose contributions are in contrast comparable and occupy only a very small fraction of the total amount of the consumed power. The by-product of this fact is the generation of excess heat, which is intolerable as it degrades the performance of the QD-SOAs and subsequently of the configuration in which they are incorporated due to the disturbance of the thermal equilibrium of their carrier dynamics and the decrease of their gain at elevated temperatures [89]-[90], while in the worst case it can cause even the damage of their QD structure. Nevertheless, the negative impact of such thermal loading can be practically compensated by employing a thermoelectric Peltier cooler and a heat sink bonded to it [91] so as to dissipate the extra heat and stabilize the temperature of the QD-SOAs. On the other hand and as far as the optically supplied power is concerned, it is worth noting that this power is not lost, for example by being totally converted into heat form. More concretely, the strong signals which are inserted in the upper and lower MZI branch and control switching also undergo amplification during their propagation in the corresponding QD-SOAs. Accordingly they exist at the MZI exit, but they do not constitute useful information. On the contrary an optical bandpass filter (OBPF) is used to reject them and select the switched weak signal that carries the logical outcome of each target Boolean function.

Last, but not least, the proposed implementation of the Boolean functions is based on the availability of the complement of one of the data signals. However in order to generate the complement of data A from the original data it is not valid to use another interferometric switch, like the QD-SOA-based MZI, since this would disagree with the concept of implementing these functions by means of only one such module. Thus it is necessary to use other methods for inverting a data signal when it contains a sequence of return-to-zero (RZ) pulses at 160 Gb/s bit rate. For this purpose there are three basic technological options which can achieve the specific goal on a signal having these characteristics. These include 1) Semiconductor optical amplifiers (SOAs) assisted by optical filtering, 2) Periodically poled lithium-niobate (PPLN) waveguides and 3) Highly nonlinear fibers (HLNF). More precisely:

1) Semiconductor Optical Amplifiers (SOAs) Assisted by Optical Filtering

This solution is distinguished from the other two in that it allows to employ a common SOA-based platform for implementing the Boolean functions, which in turn can render their design more functional, controllable, scalable and amenable to integration. It performs data logical inversion by exploiting the nonlinear effect of cross-gain modulation (XGM), which is simple and straightforward to implement in conventional or quantum-dot SOAs, according to the pump-probe configuration shown in Figure 6 [16].

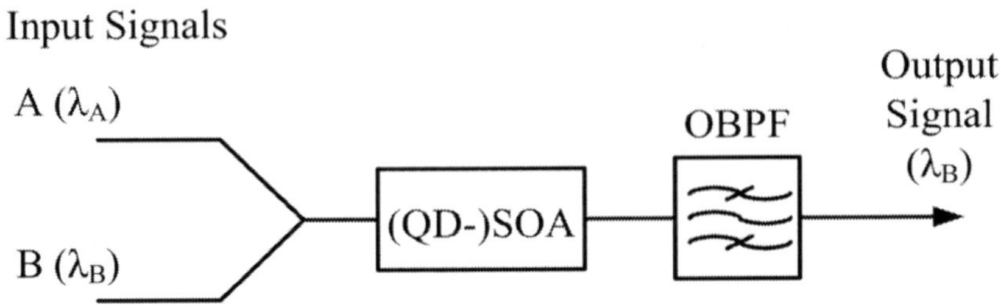

Figure 6. Generation of data A complement using (QD-) SOA assisted by optical filtering. OBPF: Optical bandpass filter.

The operation of the scheme is as follows. Two signals, the pump (A) and the probe (B) at wavelengths λ_A and λ_B, respectively, are combined and enter the SOA propagating in the same direction. The pump consists of data pulses and when it is 'on', it depletes the SOA carriers thereby dropping its gain, whilst when it is 'off' there is no saturation effect on the SOA gain. This gain modulation is experienced by the probe whose amplitude level at the output follows inversely the binary content of the pump data. Thus when the pump contains a pulse (mark) then the logical state of the emerging probe is low because it encounters a reduced SOA gain and when a bit slot of the pump is empty (space) then the logical state of the exit probe is high as it perceives a higher gain. Hence the result of this process is the complement of data signal A being mapped on signal B at wavelength λ_B, which is logically described by the function $\overline{A}B$ [92]. Now it is necessary to figure out how to enable ultrafast operation at 160 Gb/s and what form should signal B exhibit to achieve this for RZ pulses, according to the requirements for the data rate and coding, respectively, in the considered Boolean functions. The first issue can be tackled by means of employing either a bulk or a QD-SOA in combination with proper filtering using an optical bandpass filter (OBPF), as shown in the above figure, which is an approach that has been successfully adopted in the context of wavelength conversion [93]-[94]. The role of this element is that not only it blocks the pump and selects the probe but it also helps overcome the speed limitations with the associated pattern-dependent performance degradation in conventional SOAs due to their inherently slow gain recovery time [93]. Furthermore, it can improve the efficiency of XGM and accordingly the integrity of the derived signal when this nonlinear effect manifests in a QD-SOA destined for accomplishing all-optical signal processing tasks [94]. In both cases the principle of the concept relies on the fact that the pump pulses modulate not only the SOA gain but its refractive index as well. As a consequence the transmitted probe acquires a chirp whose dynamics evolve on a shorter time scale than the (QD-) SOA gain recovery. Thus by properly detuning the OBPF from the central probe wavelength it is possible to select the part of the broadened probe spectrum related to this fast chirp and enhance significantly the (QD-) SOA response. The potential of this technique for error-free and pattern-independent operation at 160 Gb/s despite the SOA recovery time being much greater than the pulses repetition period was actually demonstrated in [93]. This was achieved using RZ data pulses having a full-width at half-maximum (FWHM) pulsewidth of 1.9 ps, very close to those in our simulations, an average power of 4.8 mW and located at 1549.98 nm. The average power and the wavelength of the probe were 2.6 mW and 1560.77 nm, respectively, but regarding its form we specially comment in the following. Also the SOA was biased at 250 mA and the OBPF had a 1.4 nm bandwidth and was detuned to the lower wavelength side (blue shift) by 1.23 nm. On the other hand the quality of the output signal depends strongly on the filter's shape and wavelength offset. This means that the transfer function of the filter must be precisely designed and tailored [95], which is not an easy task, and the detuning must be sufficiently large, which however results in significant losses and reduces the optical signal to noise ratio due to the rejection of portion of the probe spectrum that contains useful information, unless we resort to sophisticated filtering solutions at the cost of increased complexity [37]. Nevertheless these tight requirements can be relaxed if a QD-SOA is used in place of the bulk SOA. Owing to the natural capability of this device to conveniently handle ultrafast data pulses it is possible to exploit the same simple mechanism of combined XGM and filtering and support an ultrafast data rate for the intended purpose with the filter playing a less critical role in terms of its shape and detuning [94]. This has been recently shown in [94] for 160 Gb/s RZ data pulses of 2.5 ps FWHM (again quite close to those in our simulations), 4.5 dBm mean power and positioned at 1543 nm. These were injected together with a 0 dBm, 1555. 3 nm probe in a 6.15-mm long SOA which consisted of columnar QDs, so that it can be made polarization-independent, was biased at 2 A and had a small signal gain and output saturation power of the order of 25 dB and 19 dBm, respectively, around 1550 nm. The conducted eye diagram measurements have revealed that by slightly blue-shifting a common type filter of 1 nm FWHM by 0.35 nm versus the probe carrier then the output signal exhibits a very high Q^2-factor of 19 dB, an improved extinction ratio, a suppressed noise level and a high power of 13 dBm, which can be controlled by a variable optical attenuator and reduced to the level required for further use. These promising results obtained for both cases suggest in turn that this (QD-) SOA-based scheme could be well employed for realizing the logical inversion of data signal A in the context of our work. To this end the second issue that must be addressed is the fact that both schemes have used a continuous-wave (CW) beam for the probe and as a consequence the output signal is a mirror image of the original RZ data in the time domain, which cannot be used as input in our QD-SOA-MZI since all driving signals must have the same polarity. What we want instead is to obtain the exact logical complement of the data signal while preserving its initial format and coding. This can be done if the CW probe is replaced by a synchronized clock signal comprising of a train of periodic pulses [96]. In this manner the XGM process takes place according to the graphical representation shown in Figure 7 [92], which eventually results in the desired logical inversion of data signal A. Therefore if the peak power of the RZ Gaussian-profile clock pulses is adjusted to match the average power, to which it is linked through the pulse duty cycle [97], of the probe used in the two reported filter-assisted XGM-based schemes [93]-[94] then the inverted data signal should be generated with similar high performance and be available in the QD-SOA-based MZI configuration that we propose to be used for implementing all-optically the referred Boolean functions.

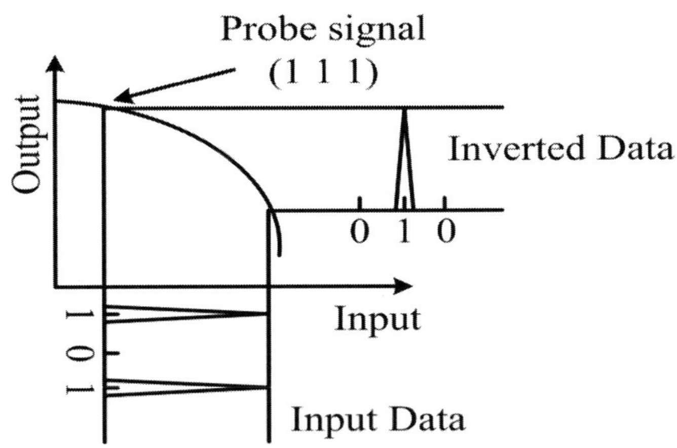

Figure 7. Principle of operation of generating the complement of data A using XGM in (QD-) SOA.

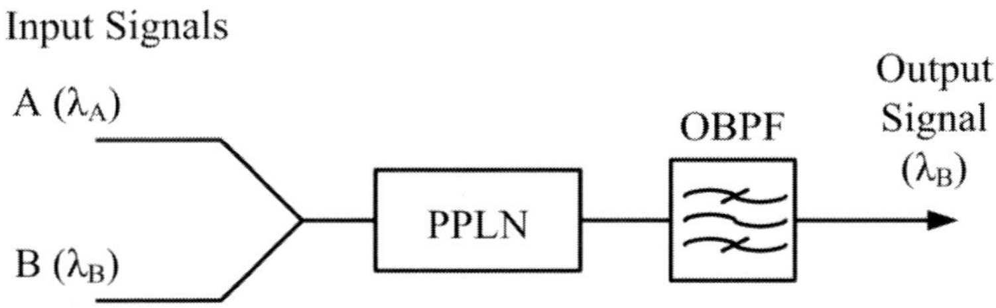

Figure 8. Generation of data A complement using PPLN waveguide. OBPF: Optical bandpass filter.

2) Periodically Poled Lithium-Niobate (PPLN) Waveguides

In this case the data logical inversion is achieved by means of a combination of sum frequency generation (SFG) and pump depletion nonlinear mechanisms, which manifest in these devices in the generic two-pump configuration whose simplified block diagram is shown in Figure 8.

The operation of the scheme is as follows. Two synchronized input signals A and B comprising of RZ data pulses and whose wavelengths λ_A and λ_B are located in the 1550 nm band are injected into the PPLN waveguide. These pumps nonlinearly interact and SFG occurs under the quasi-phase matching condition that can be set through a temperature controller attached to the waveguide [10]. During the SFG process one photon from signal A and another photon from signal B are annihilated to yield a sum-frequency (SF) photon so that signals A and B are depleted at the output. As a consequence if both A and B are at a high logical level, i.e. A = '1' and B = '1', then they are suppressed at the output, which is equivalent to a logical '0', whilst in all other combinations of logical cases where SFG does not occur they retain their initial level [10]. This essentially means that the output data bit at λ_B is '1' only when input A and B data bits are '0' and '1', respectively, which is logically expressed by the function $\overline{A}B$. Thus if signal B is a clock pulse stream so that it is continuously held to a logical '1' then we can implement the desired logical inversion for data signal A after the corresponding wavelength is selected with an OBPF. The performance of switching depends on how efficiently the involved signals are depleted, which in turn is determined by their peak power and the length of the PPLN waveguide [98]. A relevant theoretical analysis carried out to assess the impact of these parameters has specified that there is an interplay between them since as the one is decreased the other is increased and vice versa [98]. It has also been found that for optimum performance A and B should be set to the proper high power, with that of the one signal being lower than that of the other, according to the relationship $P_A = P_B \cdot \lambda_B / \lambda_A$, while in order to minimize the negative effect of the group velocity mismatch between the signals in the C-band and the converted signal at the sum frequency the length of the PPLN waveguide should be less than the walk-off length. By following these basic design rules and using approximately $P_A = 5$ W (by deduction to hyperbolic secant-shaped pulses, having 2 ps width and a practically ideal extinction ratio (ER) as in our case), $\lambda_A = 1550$ nm, $\lambda_B = 1538$ nm and a 15 mm long PPLN together with the other operating characteristics of this device detailed in [11], it has been shown through simulations that it is possible to successfully realize at 160 Gb/s the most complex of the logic operations, namely the XOR. This potential has been demonstrated experimentally as well as through bit error rate measurements [10]. Thus by keeping signal B 'on', as it was previously explained, this function can be degenerated to the simpler NOT with equally high performance in terms of the Q-factor and ER, so that it could be readily used to invert data signal A, as required in the Boolean functions implemented in our case.

3) Highly Nonlinear Fibers (HLNF)

In this case the data logical inversion is achieved by means of cross-phase modulation (XPM) [8], [99] in a fiber of high nonlinear coefficient constructed by special materials, such as chalcogenide glasses [100], according to the generic scheme shown in Figure 9.

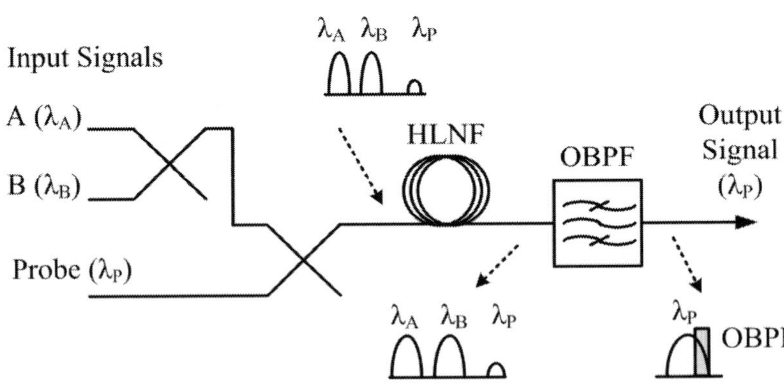

Figure 9. Generation of data A complement using HNLF. OBPF: Optical bandpass filter.

The operation of the scheme is as follows. Two strong, data-modulated signals and a weaker probe clock signal, which are located at different wavelengths, λ_A, λ_B and λ_P, respectively, in the C-band, are inserted in a highly nonlinear fiber (HNLF). Due to the fiber nonlinearity these signals interact so that the probe acquires a time-dependent nonlinear phase shift through XPM and accordingly a chirp is imposed on it given by the sum of the separate contributions induced from each data signal [8]. As a result the spectrum of the probe is broadened to an extent that depends on whether none, one or two of the data signals are present. Thus by varying their power level and selecting or removing the proper spectral slice by tuning analogously an OBPF, we can realize different logic functions. Especially for logical inversion that is of concern, only one data signal is present at the input of the fiber and the OBPF is positioned precisely at the original probe's wavelength. The data peak power must be adjusted so that it is sufficiently large to cause a high spectral shift that can be rejected by the OBPF when only one such signal is launched. Except for this requirement, the spectral separation and relative time delay of the data and the probe as well as the fiber characteristics also affect critically the performance of logical inversion. By conducting numerical simulation [8] and according to the detailed analysis and reasoning in [99], it has been predicted that it would be indeed possible to achieve logical inversion of RZ data at 160 Gb/s with a Q-factor and a contrast ratio of the order of 7 and 10 dB, respectively, for a HLNF having length 125 m, nonlinear parameter 34 $W^{-1}Km^{-1}$, loss 0.9 dB/Km, zero-dispersion wavelength 1550 nm, dispersion slope 0.03 ps/nm^2/Km, and for data/probe RZ Gaussian-shaped pulses having power and wavelength characteristics of $P_A = 200$ mW (by deduction to the 2 ps pulse width used in our simulations and assuming that the average power is the same), $\lambda_A = 1525$ nm, $P_P = 15$ mW (with the same rationale as before), $\lambda_P = 1560$ nm, with the former signal advancing the latter by 0.25 ps. Therefore if the above circuit is designed according to these realistic values of the involved parameters then when a RZ data signal A at 160 Gb/s repetition rate is inserted, a high quality logical complement of it can be generated and used to drive the corresponding input in the considered QD-SOA-based MZI Boolean functions.

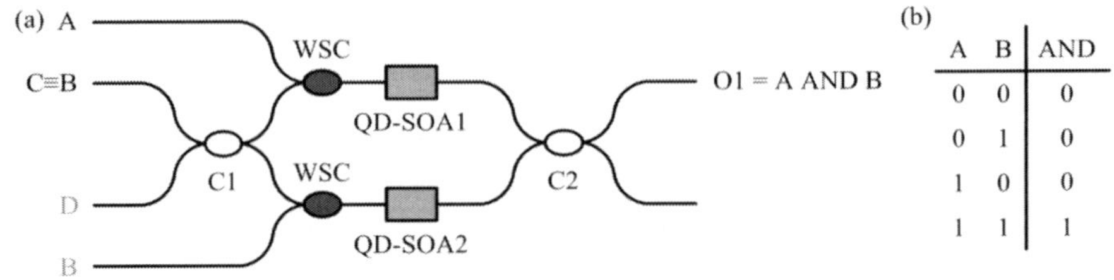

Figure 10. Modified QD-SOA-based MZI configured to execute Boolean AND operation between data A and B at output port O1. C1 and C2: 3 dB couplers. WSC: Wavelength Selective Coupler. QD-SOA: Quantum-Dot Semiconductor Optical Amplifier. (a): Block diagram. (b): Truth table.

All-Optical Boolean Functions

Selection

By properly driving the QD-SOA-based MZI configured as in Figure 5, we have managed to design several all-optical Boolean functions [101]-[103]. In this chapter we refer particularly to the AND, NAND and NOR, which have been chosen for the following reasons. First, because all are involved in the accomplishment of numerous tasks in the optical domain both in fundamental and system-oriented level. Second, because the NOR and the NAND are the universal functions by the repeated use of which it is possible to synthesize any Boolean function and hence construct any digital circuit [104]. Third, because they can be designed in a more simple, direct, affordable, efficient, scalable, cascadable and versatile way than it has been done so far [50].

Principle of Operation

The principle of operation of each target Boolean function can now be described. More specifically:

AND

Figure 10(a) shows the block diagram of the Boolean AND function. This is executed when the one data signal, A, is inserted in the modified QD-SOA-based MZI from port A, the other data signal, B, is inserted from port C, while the rest of the input ports, B and D, are idle. Then provided that the necessary conditions, which were described before in the context of the operation of the classical QD-SOA-based MZI, for the intensities and wavelengths of these two signals hold, signal A can modify the nonlinear optical properties of QD-SOA1 and induce a change on the gain and phase of signal B, while QD-SOA2 receives no such signal. Then if B = ‘0’ output at port O1 remains null no matter what the binary content of A is, because there is no input signal on which to imprint any perturbation of the initially balanced MZI and transfer it therein. This trivial situation changes when data signal B contains a pulse, i.e. B = ‘1’. In this case the logical result at O1 depends on the existence or not of a pulse in the same bit slot of data signal A. More specifically, if A = ‘0’, QD-SOA1 remains intact to the same dynamical state as QD-SOA2 so that the decomposed constituents of B perceive the same gain. Thus when they recombine at 3 dB coupler C2 they interfere destructively, which results in a space at O1. But if A = ‘1’ QD-SOA1 undergoes a radical change of its gain compared to the non-driven QD-SOA2. Consequently the copy of B in the upper MZI arm acquires via cross-phase modulation a nonlinear phase shift against its counterpart in the lower MZI arm, which eventually creates a relative phase difference between these components. If this quantity is ideally made equal to π then it is possible to maximize the amount of the power that emerges at O1 and hence the amplitude of the obtained mark. According to this mode of operation a pulse occurs at O1 if and only if a pulse is present in both signals A and B whilst no pulse appears at the specific terminal if a pulse is absent from either A or B or from both of them. In other words O1 = ‘0’ when (A, B)= (0, 0), (0, 1), (1, 0), and O1 = ‘1’ when (A, B)= (1, 1). Therefore these combinations of logical pairs and their outcome form the truth table of the Boolean AND function between A and B, as shown in Figure 10(b).

NOR

Figure 11(a) shows the block diagram of the Boolean NOR function, which is executed when the two data signals enter the modified QD-SOA-based MZI from ports A and B, respectively, while port C receives no input, i.e. C = ‘0’, and port D is driven by the complement of the data signal inserted from terminal A, i.e. $D = \overline{A}$. In fact by substituting the specified C and D in (5) and using the laws of Boolean algebra we verify that $O1 = \overline{A}(AB + \overline{A}\overline{B}) = \overline{A}\overline{B} = \overline{(A + B)}$, which by definition equals A NOR B and has the truth table shown in Figure 11(b). Note that A and B are introduced into a different QD-SOA device, which offers significant practical benefits. First, we can use the same wavelength for A and B in order to discriminate them from D inside the MZI. This allows to minimize the number of laser sources that are required for providing these signals to two, which is an important saving in hardware resources [105]. Concurrently it enables to execute the NOR function logic with reduced cost and complexity. Second, since each QD-SOA is now driven with only one strong signal, we can control and keep the level of light intensity going into it to a reasonable level. In this manner we can lessen the burden that would be imposed on the QD-SOA gain dynamics by the excessive sum of the power of two strong data signals, like A and B, if were instead launched simultaneously into it, which could impair the QD-SOA performance and in particular its ultrafast capability [106].

(a)

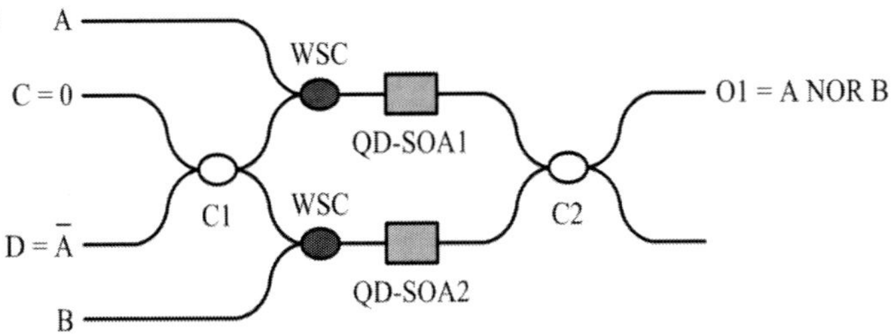

(b)

A	B	NOR
0	0	1
0	1	0
1	0	0
1	1	0

Figure 11. Modified QD-SOA-based MZI configured to execute Boolean NOR operation between data A and B at output port O1. C1 and C2: 3 dB couplers. WSC: Wavelength Selective Coupler. QD-SOA: Quantum-Dot Semiconductor Optical Amplifier. (a): Block diagram. (b): Truth table.

(a)

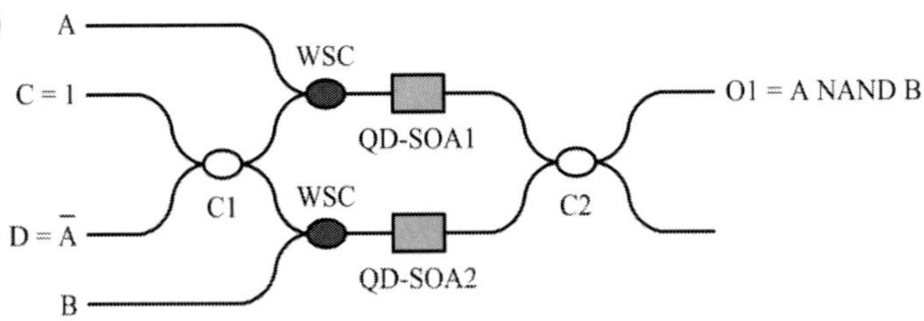

(b)

A	B	NAND
0	0	1
0	1	1
1	0	1
1	1	0

Figure 12. Modified QD-SOA-based MZI configured to execute Boolean NAND operation between data A and B at output port O1. C1 and C2: 3 dB couplers. WSC: Wavelength Selective Coupler. QD-SOA: Quantum-Dot Semiconductor Optical Amplifier. (a): Block diagram. (b): Truth table.

NAND

Figure 12(a) shows the block diagram of the Boolean NAND function, which is executed when the two data signals enter the modified QD-SOA-based MZI from ports A and B, respectively, while port C is driven by a clock, i.e. C = '1', and port D by the complement of the data signal A, i.e. $D=\overline{A}$. This can be readily verified with the help of Boolean algebra by substituting the specified C and D in (5), which yields indeed $O1=\overline{A}+A\overline{B}=\overline{A}+\overline{B}=\overline{AB}$ whose truth table is shown in Figure 12(b). The required complementary data can be generated from the original data by employing one of the technological options described in the last part of the 'Real Implementation Issues' section. It is noteworthy that the proposed scheme employs a single QD-SOA-based MZI switch. This enables to realize the NAND function in a straightforward and accordingly more affordable manner without the limitations imposed by cascading two switching units, i.e. an AND and a NOT, for achieving the same logical operation [65], [72].

OR

Finally, using a similar rational as for the implementation for the NOR and NAND, Figure 13(a) shows the block diagram of the OR function. This is formed when the two data signals are introduced in the modified QD-SOA-based MZI from ports A and B, respectively, while port C is driven by a clock, i.e. C = '1', and port D by a signal containing the same binary information as that inserted from port A, i.e. D = A, so as to realize the truth table shown in Figure 13(b).

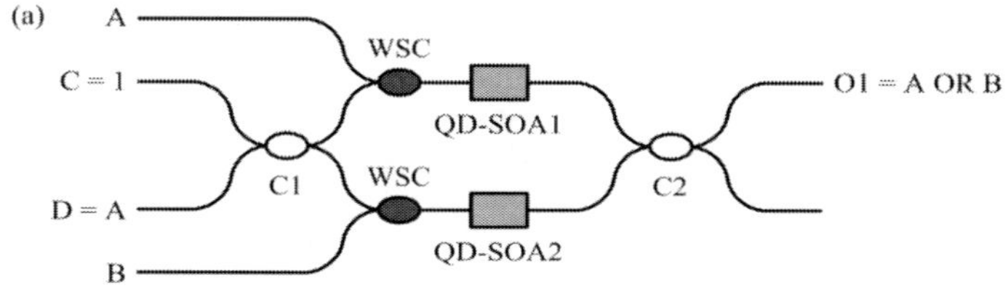

(b)

A	B	OR
0	0	0
0	1	1
1	0	1
1	1	1

Figure 13. Modified QD-SOA-based MZI configured to execute Boolean OR operation between data A and B at output port O1. C1 and C2: 3 dB couplers. WSC: Wavelength Selective Coupler. QD-SOA: Quantum-Dot Semiconductor Optical Amplifier. (a): Block diagram. (b): Truth table.

Note that in order to be able to discriminate inside the MZI the signals launched from terminals A and D, a wavelength converter must be placed in the path of data A. A possible way of doing this at 160 Gb/s in the context of having a unified SOA-based platform is by means of cross-gain modulation in a SOA assisted by a properly detuned filter and an optical delay interferometer in a pump-probe scheme [93], where the pump consists of the data pulses and the probe of a continuous wave beam at the new wavelength.

Critical Operating Parameters

Given the important and multi-lateral role of these functions we have addressed their design so that they can be exploited in the wide range of applications destined to serve at excessively increased single channel data rates. For this purpose we have thoroughly investigated and assessed the influence of critical operating parameters on defined performance metrics. These critical operating parameters are selected according to the launched data and QD-SOA important characteristics and their significance and role for the operation of the QD-SOAs and subsequently for the MZI in which the latter are incorporated. Thus they include the peak power of the launched data pulses, P_{peak}, and the QD-SOAs small signal gain, G_{ss}, current density, J, and electron relaxation time from the ES to the GS, τ_{21}.

More specifically, the peak data power determines the degree of the QD-SOAs saturation under pulsed mode of operation with respect to the 3 dB input saturation power. Subsequently it determines the reduction of the gains of the QD-SOAs whose relative ratio, $\frac{G_1}{G_2}$, between the MZI arms is responsible for inducing the phase difference, $\Delta\varphi = -\frac{\alpha_{LEF}}{2}\ln\left(\frac{G_1}{G_2}\right)$, between the copies of the switched signals C and D. Although a high peak power may be required to achieve the desired Boolean operation, yet it must also be reasonable so that it can be provided from commercially available erbium doped fiber amplifiers.

Furthermore, the QD-SOAs small signal gain parameter determines the extent that the gain of each QD-SOA is dropped from the unsaturated level [51] and subsequently the amount of the differential gain created between the MZI arms. This parameter primarily depends on the maximum modal gain, which scales with the number, surface density and effective thickness of the QD layers, and the length of the active waveguide, which influences the amount of phase shift that accumulates in each QD-SOA [22]. Thus a balance must be maintained between these factors so that the construction of the QD-SOA is practically feasible while the necessary phase change can be created through the proper change of the QD-SOA optical properties.

Moreover, the current density determines the amount of supplied carriers that are available for gain recovery as well as the saturation power of each QD-SOA [107]. A crucial point regarding the selection of this parameter is that the equivalent bias current should be supplied by the commercially available current sources while not being prohibitive for the practical use of QD-SOAs [106].

Finally, the electron relaxation time from the ES to the GS determines how fast the carrier replenishment and hence the QD-SOA gain recovery takes place, which affects decisively the capability of the QD-SOAs to timely respond to ultrafast launched data. The values of this parameter must comply from a practical perspective with the time scale at which the transition between the ES and GS occurs based on pump-probe measurements of the time-resolved response of QD-SOAs [108]-[109], so that they are technologically achievable.

Performance Metrics

In order to evaluate the performance of each Boolean function, we must choose the appropriate metrics. This can be done by considering those which are standardized for describing quantitatively the performance of lightwave digital circuits and systems. These include the Q-factor, the amplitude modulation (AM) and the extinction ratio (ER). First, the Q-factor is defined as [110]

$$Q = \frac{\overline{P}_1 - \overline{P}_0}{\sigma_1 + \sigma_0} \tag{10}$$

where $\overline{P}_1, \overline{P}_0$ and σ_1, σ_0 are the mean and the standard deviation of the peak power of the marks and spaces, respectively, at output O1. For error-free operation, which is quantified by a bit error rate of less than 10^{-9}, the Q-factor must at least equal to six [110], which therefore defines the lower bound that must be exceeded in an all-optical Boolean operation of the modified QD-SOA-based MZI switch.

Second, the amplitude modulation is defined as [23]

$$AM = 10 \log\left(\frac{P_1^{max}}{P_1^{min}}\right) \quad (11)$$

where P_1^{max} and P_1^{min} are the maximum and minimum peak power, respectively, among the '1's that are contained in the sequence that is switched at port O1 of the modified QD-SOA-based MZI as a result of the Boolean operation executed between data A and B. The AM expresses the degree of uniformity of the marks and if it is lower than 1 dB then the amplitude fluctuations of these bits and the concomitant pattern effect imposed on them through the switching process can be kept at an acceptably low level [111].

Third, the extinction ratio is defined as [23]

$$ER\ (dB) = 10\log\left(\frac{P_{min}^1}{P_{max}^0}\right) \quad (12)$$

where P_{min}^1 and P_{max}^0 are the minimum and maximum peak power of the switched marks and spaces. The ER must be over 10 dB [87] so that the '1's can be clearly distinguished from the '0's, which is a basic prerequisite for executing the Boolean functions with high performance. Furthermore it defines the degree of uniformity and opening of the so-called pseudo-eye diagrams (PEDs) [112].

Design Results

In order to examine whether the considered Boolean functions can be executed in the QD-SOA-based MZI configuration that corresponds to each one of them, their performance must be evaluated against the metrics defined above. The selection of the most appropriate metric for each Boolean function depends on the special nature of its logical outcome in the corresponding truth table, in combination with its principle of all-optical implementation using the modified QD-SOA-based MZI. More specifically:

a) For the AND, we perceive that a mark is produced only for one logical combination, i.e. (A, B) = (1, 1). Furthermore, we realize that in the all-optical implementation the maximum number of marks will occur when, according to the described principle of operation, the patterns of data inputs A and B are exactly the same and synchronized to coincide. Then in this case the number of output marks equals the number of marks within the pseudorandom binary sequence (PRBS), which is 2^{n-1} for a word length of $2^n - 1$. Concurrently, however, there will also be $2^{n-1} - 1$ spaces [113]. This means that in the specific, extreme though realistic scenario, there will nearly be a balanced contribution from ones and zeros. Hence this situation, which can be rather demanding especially when the launched data are as fast as of the order of sub-Tb/s and additionally the input zeros have a finite extinction ratio [101], dictates to take into account the peak power of all these bits, which by definition can be done by calculating the Q-factor.

b) For the NOR, we note that when (A, B) = (1, 0) and (1, 1), then $D = \overline{A} = 0$ and accordingly $P_D(t) = 0$, so that $P_{O1} = 0$ from (6), which corresponds to fully extinguished spaces. When (A, B) = (0, 0), then $G_1(t) \approx G_2(t)$, and $D = \overline{A} = 1$, so that P_{O1} is maximized, which corresponds to marks of uniform amplitude. Finally, the impact of (A, B) = (0, 1) lies in between these cases. Therefore the reliable assessment of the performance of this function requires to compare the level of the marks and the spaces that occur from the logical pairs (A, B) = (0, 0) and (A, B) = (0, 1). Thus the most suitable metric for this purpose is the extinction ratio.

c) For the NAND and OR, we observe that of the four logical results that are possible for these functions, three concern a mark and one a space. In particular the space is obtained when both data signals A and B are present or absent, respectively. This means that the dynamical behavior of the QD-SOAs in the two MZI arms is either affected in a common manner or not perturbed at all. As a practical consequence the resultant spaces are fully extinguished and can be clearly distinguished from the marks. This fact rationally leads us to focus on the marks and their possible peak-to-peak wandering whose extent can be checked with the help of the amplitude modulation.

Then the impact of the identified critical operating parameters on the performance metric employed for each Boolean function has been investigated and assessed through the thorough observation and analytical interpretation of numerically derived simulation curves. This procedure has allowed us to extract rules for the implementation of each target Boolean function, which if followed then these functions can be executed with logical correctness and high quality. Thus in the following we provide concise instructions for the design of the modified QD-SOA-based MZI so that each Boolean function can be executed all-optically, as shown in the form of frames below:

AND

For this Boolean function the requirements that have been derived for the critical parameters are $4.4\,\text{dBm} \leq P_{\text{peak}} \leq 9.2\,\text{dBm}$, $10.8\,\text{dB} \leq G_{ss} \leq 14.8\,\text{dB}$, $J \geq 0.9\,\text{kA/cm}^2$ and $0.3\,\text{ps} \leq \tau_{21} \leq 1.47\,\text{ps}$. Thus by following these guidelines and using the combination of values $P_{\text{peak}} = 9$ dBm, $G_{ss} = 14$ dB, $J = 1.5$ kA/cm^2 and $\tau_{21} = 0.6$ ps, an exceptionally high Q-factor (39.8) can be achieved. This is reflected on the profile of the pulses obtained for the representative data patterns A = '1011011110110001' and B = '1001010101111111', as shown in Figure 14.

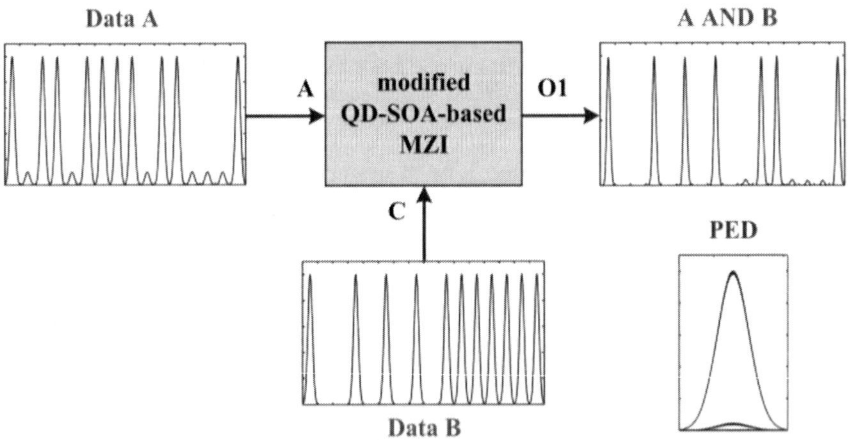

Figure 14. Results for Boolean AND function all-optical implementation with QD-SOA-based MZI. Top-right: Logical outcome. Bottom-right: Pseudo-eye diagram (PED).

Indeed the logical outcome complies with the truth table of the specific Boolean function, since a mark is produced only when the binary content of A and B is identical while a space occurs in all other cases. Furthermore the zeros are strongly suppressed, since their peak level is at least 13.3 dB lower than that of the marks. At the same time the marks are sufficiently balanced and the difference between their peak amplitudes is only 0.1 dB. The good quality of the AND function is also confirmed by the excellent form of the PED, which has been plotted for a 127 bit-long data stream. This visual quality indicator is obtained by superimposing the marks and spaces in the data stream of interest on top of each other and displaying them within a single bit period. Then we can see that the spaces have acquired a single and suppressed border while the envelope of the marks is uniform since their individual traces almost coincide having a hardly discernible difference between them. Thus the PED is clear and open, which ensures that the AND function can be realized all-optically without errors.

NOR

For this Boolean function the requirements that have been derived for the critical parameters are $10.1\ \text{dBm} \leq P_{peak} \leq 12.7\ \text{dBm}$, $17.9\ \text{dB} \leq G_{ss} \leq 20.3\ \text{dB}$, $J \geq 1.4\ \text{kA/cm}^2$ and $0.18\ \text{ps} \leq \tau_{21} \leq 0.4\ \text{ps}$. Thus making the suitable selection $P_{peak} = 10.1\ \text{dBm}$, $G_{ss} = 19\ \text{dB}$, $J = 3\ \text{kA/cm}^2$ and $\tau_{21} = 0.22\text{ps}$, the NOR function can be executed between the indicative data patterns A = '0011100010010011' and B = '1000100100110110' with an adequate ER = 13.1 dB. This is reflected on the profile of the obtained pulses shown in Figure 15. In fact the largest amount of power emerges at the bit slots of A and B that are both empty, as expected according to the truth table of this Boolean function. Furthermore, the obtained marks have the same height while the spaces are strongly suppressed when they originate from (A, B) = (0, 1) and are fully extinguished when they originate from (A, B) = (1, 0) or (1, 1). These observations designate that the outcome of the NOR function is logically correct and of high quality. Moreover, in direct analogy to these results, the corresponding PED is open, with the marks having a negligible peak difference and the spaces a single envelope.

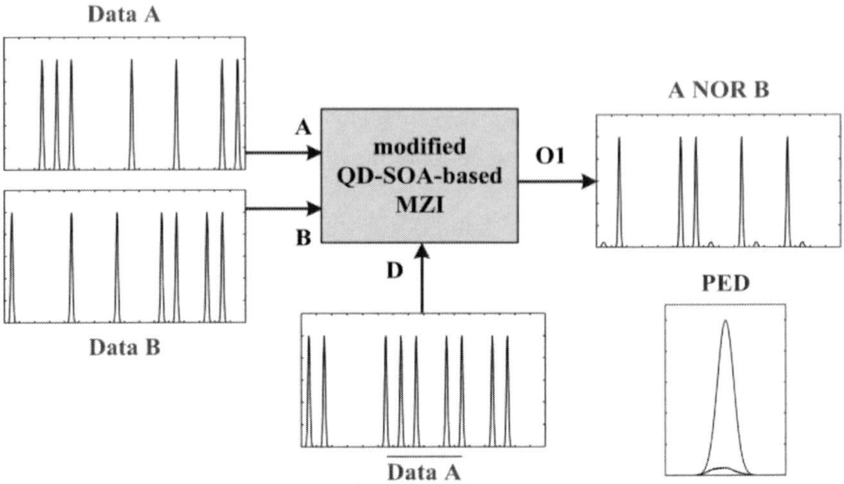

Figure 15. Results for Boolean NOR function all-optical implementation with QD-SOA-based MZI. Top-right: Logical outcome. Bottom-right: Pseudo-eye diagram (PED).

NAND

For this Boolean function the requirements that have been derived for the critical parameters are $8.65\ \text{dBm} \leq P_{peak} \leq 12.7\ \text{dBm}$, $16.35\ \text{dB} \leq G_{ss} \leq 20.4\ \text{dB}$, $J \geq 1.5\ \text{kA/cm}^2$ and $0.15\ \text{ps} \leq \tau_{21} \leq 0.45\ \text{ps}$. Thus using the combination of values $P_{peak} = 10.1\ \text{dBm}$, $G_{ss} = 19\ \text{dB}$, $J = 3\ \text{kA/cm}^2$ and $\tau_{21} = 0.22\text{ps}$, which fall within the specified boundaries, the NAND function can be realized with high performance, as shown in Figure 16, between the indicative data patterns A = '0011100010010011' and B = '1000100100110110'. In fact pulses emerge only at the bit slots of A and B that are not simultaneously occupied by '1' or equivalently if just one of these inputs is in the '0' logical state, and are extinguished when both A and B are '1'. This proves that the NAND operation is correctly executed according to its truth table. The same holds if the resultant sequence is compared against the expected one, $\overline{AB}$ = '1111011111101101'. Moreover, despite the non-uniformity among the output marks, the quality of the logical outcome is more than adequate since the difference in their amplitudes is strongly suppressed to

0.23 dB. The existing pattern dependence is inevitable and cannot be fully eliminated due to the special nature of switching, which in the particular case of the specific function demands to combat the amplitude fluctuations of marks that correspond to a high percentage, 75%, of the output logical possibilities. Nevertheless, this is a fundamental rather than a conceptual difficulty, which is compensated by the combination of the QD-SOAs technological advantages and the practical benefits offered by the modified MZI architecture. The potential of the scheme for Boolean NAND functionality is further supported by the PED, whose sub-envelopes have a small vertical distance from the top owing to the achievement of low AM.

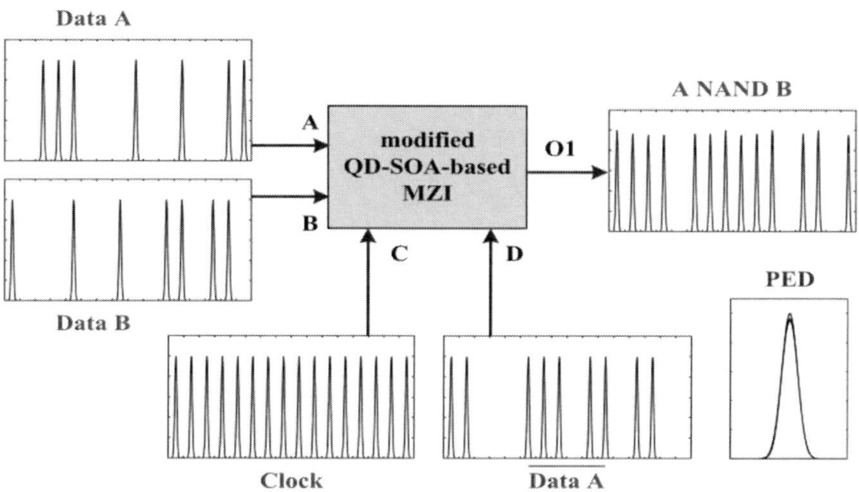

Figure 16. Results for Boolean NAND function all-optical implementation with QD-SOA-based MZI. Top-right: Logical outcome. Bottom-right: Pseudo-eye diagram (PED).

OR

Finally, the design requirements for the execution of the OR function are covered by those for the other Boolean functions. Consequently this operation can be achieved between data patterns A = '0011100010010011' and B = '1000100100110110' with a low amplitude modulation AM = 0.26 dB at output O1 of the modified QD-SOA-based MZI. The obtained logical result, which is shown in Figure 17, complies with the OR truth table, while the quality of the PED is as high as that of the logical outcome.

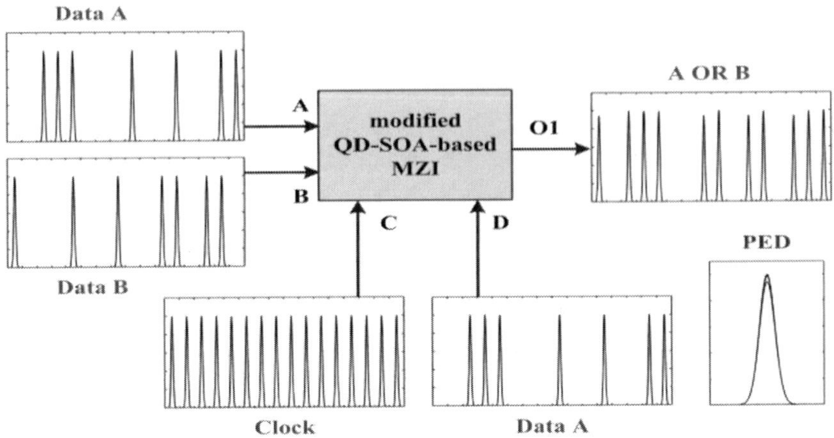

Figure 17. Results for Boolean OR function all-optical implementation with QD-SOA-based MZI. Top-right: Logical outcome. Bottom-right: Pseudo-eye diagram (PED).

APPLICATION: ULTRAFAST ALL-OPTICAL 4-BIT PARITY GENERATOR & CHECKER

The scope of designing and executing all-optically Boolean functions is broader than using them only as stand-alone discrete logic entities. In fact the ultimate goal is to exploit them for constructing more complex circuits of enhanced combinational and sequential functionality. For this purpose theoretical results have been reported on all-optical half-adder [114], pseudorandom binary generator [115] and latches [116] using the QD-SOA-based MZI as switching module. Recently the set of these demonstrations was extended to include an ultrafast 4-bit parity generator and checker, as concisely described in the following.

The architecture of the proposed circuit is shown in Figure 18(a). Four XOR gates are cascaded in a modular manner. The XOR gates are implemented according to the described principle of operation of the classical QD-SOA-based MZI in dual rail switching mode. The binary information to undergo parity generation and checking is a 4-bit word, which consists of four input digits, X_1, X_2, X_3 and X_4. Initially X_1 and X_2 are inserted in the first XOR gate, XOR 1, which hence produces the logical result $X_2 \oplus X_1$. The output O_1 is afterwards fed to one data input of the subsequent XOR gate,

XOR 2, whose other data input receives the third word bit, X_3. Similarly output O_2 of XOR 2 is connected to one data input of XOR 3 whose output O_3 is therefore the modulo-2 sum of all word bits until that point, i.e. $O_3 = X_4 \oplus X_3 \oplus X_2 \oplus X_1$. Now in order for the circuit in Figure 18(a) to perform parity generation and checking, another XOR gate, XOR 4, is added in the setup. This is driven by output O_3 of the penultimate XOR gate and another input, P'. The logical state of this input determines whether the circuit will function as parity generator or as a parity checker.

(a)

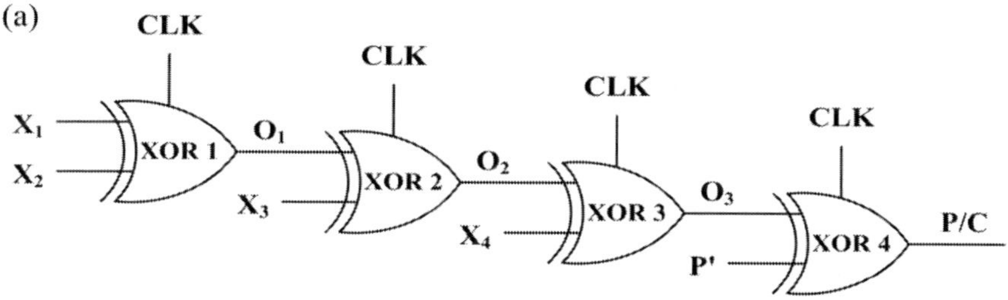

(b)

4-bit word				Parity bit
X_4	X_3	X_2	X_1	P
0	0	0	0	0
0	0	0	1	1
0	0	1	0	1
0	0	1	1	0
0	1	0	0	1
0	1	0	1	0
0	1	1	0	0
0	1	1	1	1
1	0	0	0	1
1	0	0	1	0
1	0	1	0	0
1	0	1	1	1
1	1	0	0	0
1	1	0	1	1
1	1	1	0	1
1	1	1	1	0

4-bit word + Parity bit					Parity check	
X_4	X_3	X_2	X_1	P	C	
0	0	0	0	0	0	
0	0	0	1	1	0	
0	**<u>1</u>**	1	0	1	1	error
0	0	1	1	0	0	
0	1	0	0	1	0	
<u>1</u>	1	0	1	0	1	error
0	1	1	0	0	0	
0	1	1	1	1	0	
1	0	0	**<u>1</u>**	1	1	error
1	0	0	1	0	0	
1	0	1	0	0	0	
1	**<u>1</u>**	1	1	1	1	error
1	1	0	0	0	0	
1	1	0	1	1	0	
1	1	1	0	1	0	
1	1	1	1	0	0	

Figure 18. Proposed all-optical 4-bit parity processing circuit. (a) Layout and (b) truth table for 4-bit parity generation (left-hand side) and checking (right-hand side). The underlined bold marks denote the occurrence of an error.

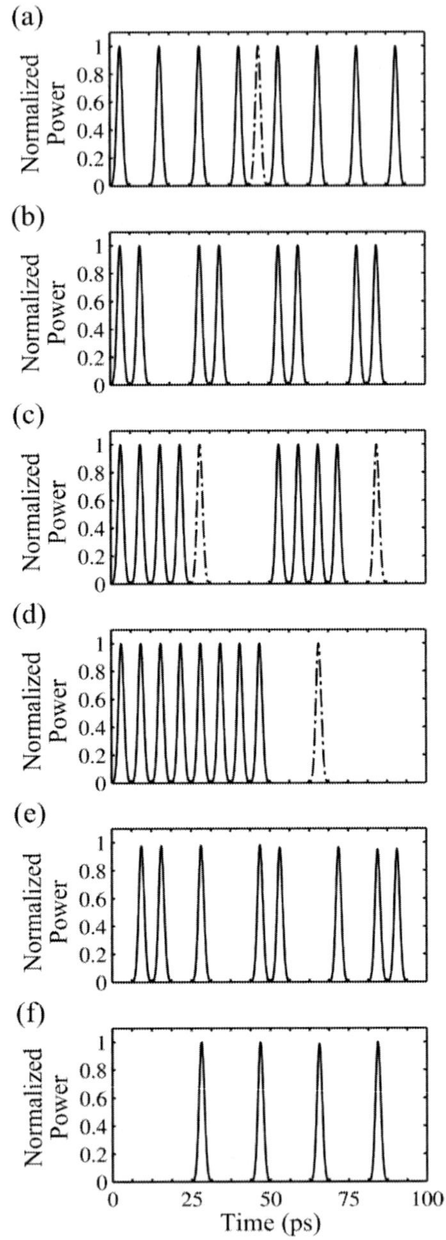

Figure 19. Results for all-optical 4-bit even parity generation (e) and checking (f), which is realized on data inputs X1 (a), X2 (b), X3 (c) and X4 (d), where for (e) $P' =$ '000...0' and for (f) $P' = P$ taken from (e).

More specifically, if $P' =$ ‘0’ the output of the rightmost XOR gate is

$$P = P' \oplus (X_4 \oplus X_3 \oplus X_2 \oplus X_1) = X_4 \oplus X_3 \oplus X_2 \oplus X_1 \qquad (13)$$

From the truth table of this expression in the left-hand side of Figure 18(b) it can be seen that for all the possible logical combinations of X_1, X_2, X_3 and X_4, P becomes ‘1’ when the number of ‘1’s in each line is odd and ‘0’ when the corresponding number is even. This means that with the inclusion of P the total number of ‘1’s in the expanded sequence, X_1, X_2, X_3, X_4, P, is always even and so the circuit of Figure 18(a) is a 4-bit even parity generator [104]. Thus when the information bits are forwarded together with the redundant bit to their destination the total number of ‘1’s at the receiving point must remain even so as to ensure correct communication. The satisfaction of this condition can be checked by using again the same circuit of Figure 18(a) but this time replacing P' with the parity bit generated according to (13), i.e. $P' \equiv P$. The new output is then

$$C = P \oplus X_4 \oplus X_3 \oplus X_2 \oplus X_1 \qquad (14)$$

and the truth table of this expression is given in the right-hand side of Figure 18(b). Now if C = ‘1’ then the total number of ‘1’s is no longer even, which indicates that during transmission from a position to another an error has occurred (denoted by underlined bold marks). In contrast if C = ‘0’ there is no such problem since even parity is preserved [104]. The circuit can also support odd parity processing. For odd parity generation we must set $P' =$‘1’ so that $P = \overline{X_4 \oplus X_3 \oplus X_2 \oplus X_1}$, while for odd parity checking we require C = ‘1’ for error-free operation.

By conducting numerical simulation we have specified the requirements for the critical parameters in order for the all-optical 4-bit parity generator and checker to function correctly while ensure an amplitude modulation AM < 1 dB at its output. These dictate that the peak power of the data signals at the two available inputs of each XOR gate must be less than 14 dBm, the QD-SOAs small signal gain must not surpass 17.5 dB, the QD-SOAs current density must exceed 1.8 kA/cm^2, and the QD-SOAs electron relaxation time from the WL to the ES must not be over 2.5 ps. Thus by following these guidelines and using the values 7 dBm, 16 dB, 1.8 kA/cm^2 and 1.5 ps, respectively, the parity processor can execute its target operations on 4-bit binary words with logical correctness and acceptable quality. This capability is verified by the results obtained for the outcome, P, of even parity generation, which are shown in Figure 19(e). For visual convenience, the exemplary 16-bit long data patterns $X_1 =$'1010101010101010' (Figure 19(a)), $X_2 =$ '1100110011001100' (Figure 19(b)), $X_3 =$'1111000011110000' (Figure 19(c)), $X_4 =$'1111111100000000' (Figure 19(d)) are also portrayed, while $X_5 \equiv P' =$'000...0'.

By combining the bits of X_1, X_2, X_3 and X_4 that are located in the same pulse position, different 4-bit words are formed. Then from Figure 19(e) it can be confirmed that a space is produced at output P when the number of marks in each such binary word is even and a mark is produced therein when the number of marks is odd. These logical outcomes coincide with the truth table of even parity generation in the left-hand side of Figure 18(b), which proves that the parity bits are generated correctly by the all-optical parity processing scheme. Furthermore Figure 19(f) depicts the outcome, C, of even parity checking, where P' coincides with P taken from Figure 19(e). In this case we have used characteristic combinations of X_1, X_2, X_3 and X_4 in which an error has occurred in some bit position. Without loss of generality we have assumed that this error happens when a space is impaired, as indicated by the dashed-dotted envelope. Thus at the bit slots where such an error has been detected, a mark appears in Figure 19(f), which is consistent with the principle of operation of the circuit. It is also noteworthy that in all tested scenarios the spaces are fully suppressed while the marks have nearly the same amplitude. This observations are quantified by AM = 0.22 dB for even parity generation and AM = 0.05 dB for even parity checking. Finally, equally good operation and performance is achieved for odd parity generation and checking.

CONCLUSION

The all-optical implementation of a set of fundamental Boolean functions has been proposed and demonstrated. The core photonic switching unit has been provided by a modified Mach-Zehnder interferometer structure that takes advantage of the unique properties of quantum-dot semiconductor optical amplifiers. The operation of basic and universal Boolean functions has been analyzed, assessed and optimized through numerical modeling of the QD-SOAs, both as stand-alone elements and incorporated in the MZI, in response to excessively fast input data. This design procedure has allowed to specify in a systematic manner the conditions that must be satisfied for the best selection of the critical performance parameters for the QD-SOA-based MZI modules which are properly configured and driven to realize each Boolean function. The successful prediction of the feasibility of these functions by controlling light with light can have a positive impact on a) the continuous evolution of QD materials and devices towards even commercial exploitation, b) the construction of more complex optical circuits and subsystems, c) the support of enhanced combinational and sequential functionality at the optical level, d) the development of larger scale arrays of photonic switches and architectures of multiple logical stages, e) the accomplishment of the vision of optical digital computing, f) the enrichment of enabling technologies that motivate optical signal processing, and, overall, g) the strengthening of the information society through the global penetration of fiber optics technology and its multilateral benefits.

References

[1] Bogoni, A; Potì, L; Ghelfi, P; Scaffardi, M; Porzi, C; Ponzini, F; Meloni, G; Berrettini, G; Malacarne A; Prati, G. "OTDM-based optical communications networks at 160 Gbit/s and beyond", *Optical Fiber Technology*, Vol. 13, No. 1, 1-12, 2007.

[2] Mulvad, HCH; Galili, M; Oxenløwe, LK; Hu, H; Clausen, AT; Lensen, JB; Peucheret, C; Jeppesen, P. "Demonstration of 5.1 Tbit/s data capacity on a single-wavelength channel", *Optics Express*, Vol. 18, No. 2, 1438-1443, 2010.

[3] Willner, AE; Khaleghi, S; Chitgarha, MR; Yilmaz, OF. "All-optical signal processing", *Journal of Lightwave Technology*, Vol. 32, No. 4, 660-680, 2014.

[4] Karim MA; Awwal, AAS. *Optical Computing: An Introduction*, Wiley, New York, 1992.

[5] Singh, P; Tripathi, D. Kr; Jaiswal S; Dixit, HK. "All-optical logic gates: Designs, classification and comparison", *Advances in Optical Technologies*, Vol. 2014, art. No. 275083, 1-13, 2014.

[6] Kaminow, IP; Li T; Willner, AE. *Optical Fiber Telecommunications V, B: Systems and Networks*, ch. 2, Academic Press, 2008.

[7] Kang, I; Rasras, M; Buhl, L; Dinu, M; Cabot, S; Cappuzzo, M; Gomez, LT; Chen, YF; Patel, SS; Dutta, N; Piccirilli, A; Jaques, J; Giles, CR. "All-optical XOR and XNOR operations at 86.4 Gb/s using a pair of semiconductor optical amplifier Mach-Zehnder interferometers", Optics Express, Vol. 17, No. 21, 19062-19066, 2009.

[8] Bogris, A; Velanas P; Syvridis, D. "Numerical investigation of a 160-Gb/s reconfigurable photonic logic gate based on cross-phase modulation in fibers", *IEEE Photonics Technology Letters*, Vol. 19, No. 6, 402-404, 2007.

[9] Miyoshi, Y; Ikeda, K; Tobioka, H; Inoue, T; Namiki, S; Kitayama, K. "Ultrafast all-optical logic gate using a nonlinear optical loop mirror based multi-period transfer function", *Optics Express*, Vol. 16, No. 4, 2570-2577, 2008.

[10] Bogoni, A; Wu, X; Fazal, I; Willner, AE. "160 Gb/s time-domain channel extraction/insertion and all-optical logic operations exploiting a single PPLN waveguide", *Journal of Lightwave Technology*, Vol. 27, No. 19, 4221-4227, 2009.

[11] Wang, J; Sun, J; Sun, Q; Zhang X; Huang, D. "Simple realization of all-optical high-speed (40, 80 and 160 Gb s^{-1}) XOR and OR logic gates using $LiNbO_3$ waveguides", *Journal of Optics A: Pure and Applied Optics*, Vol. 9, 811-819, 2007.

[12] Lee, YL; Yu, B-A; Eom, TJ; Shin, W; Jung, C; Noh, Y-C; Lee, J; Ko, D-K. "All-optical AND and NAND gates based on cascaded second-order nonlinear processes in a Ti-diffused periodically poled $LiNbO_3$ waveguide", *Optics Express*, Vol. 14, No. 7, 2776–2782, 2006.

[13] Drummond, MV; Reis, JD; Nogueira, RN; Monteiro, PP; Teixeira, AL; Shinada, S; Wada, N; Ito, H. "Error-free wavelength conversion at 160 Gbit/s in PPLN waveguide at room temperature", *Electronics Letters*, Vol. 45, No. 22, 1135-1137, 2009.

[14] Dagens, B; Labrousse, A; Brenot, R; Lavigne, B; Renaud, M. "SOA-based devices for all-optical signal processing", *Proceedings Optical Fiber Communications (OFC)*, ThX1, 582-583, 2003.

[15] Patel, NS; Hall, KL; Rauschenbach, KA. "Interferometric all-optical switches for ultrafast signal processing", *Applied Optics*, Vol. 37, No. 14, 2831–2842, 1998.

[16] Mørk, J; Nielsen, ML; Berg, TW. "The dynamics of semiconductor optical amplifiers: Modeling and applications", *Optics & Photonics News*, Vol. 14, No. 7, 42-48, 2003.

[17] Schares, L; Schubert, C; Schmidt, C; Weber, H. G; Occhi, L; Guekos, G. "Phase dynamics of semiconductor optical amplifiers at 10-40 GHz", *IEEE Journal of Quantum Electronics*, Vol. 39, No. 11, 1394-1408, 2003.

[18] Giller, R; Manning, RJ; Talli, G; Webb, RP; "Analysis of the dimensional dependence of semiconductor optical amplifier recovery speed", *Optics Express*, Vol. 15, No. 4, 1773-1782, 2007.

[19] Xia, G-Q; Wu, Z-M; Lin, G-R. "Rising and falling time of amplified picosecond optical pulses by semiconductor optical amplifiers", *Optics Communications*, Vol. 227, No. 1-3, 165-170, 2003.

[20] Bischoff, S; Buxens, A; Fischer, S; Dülk, M; Clausen, AT; Poulsen, H. N; Mørk, J. "Comparison of all-optical co-and counter-propagating high-speed signal processing in SOA-based Mach-Zehnder interferometers", *Optical and Quantum Electronics*, Vol. 33, No. 7-10, 907-926, 2001.

[21] Diez, S; Hilliger, E; Kroh, M; Schmidt, C; Schubert, C; Weber, HG; Occhi, L; Schares, L; Guekos, G; Oxenloewe, LK. "Optimization of SOA-based Sagnac-interferometer switches for demultiplexing to 10 and 40 Gb/s", *Optics Communications*, Vol. 189, No. 4-6, 241-249, 2001.

[22] Ueno, Y; Nakamura, S; Tajima, K. "Nonlinear phase shifts induced by semiconductor optical amplifiers with control pulses at repetition frequencies in the 40-160-GHz range for use in ultrahigh-speed all-optical signal processing", *Journal of Optical Society of America B*, Vol. 19, No. 11, 2573-2589, 2002.

[23] Zoiros, KE; Papadopoulos, G; Houbavlis, T; Kanellos, GT. "Theoretical analysis and performance investigation of ultrafast all-optical Boolean XOR gate with semiconductor optical amplifier-assisted Sagnac interferometer", *Optics Communications, Vol.* 258, No. 2, 114-134, 2006.

[24] Inohara, R; Nishimura, K; Tsurusawa, M; Usami, M. "Experimental analysis of cross-phase modulation and cross-gain modulation in SOA-injecting CW assist light", *IEEE Photonics Technology Letters*, Vol. 15, No.9, 1192-1194, 2003.

[25] Wang, H; Wu, J; Lin, J. "Spectral characteristics of optical pulse amplification in SOA under assist light injection", *Journal of Lightwave Technology*, Vol. 23, No. 9, 2761-2771.

[26] Manning, RJ; Sherlock, G. "Recovery of a π phase shift in ~12.5 ps in a semiconductor laser amplifier", *Electronics Letters*, Vol. 31, No. 4, 307-308, 1995.

[27] Pleros, N; Kanellos, GT; Bintjas, C; Hatziefremidis, A; Avramopoulos, H. "Optical power limiter using a saturated SOA-based interferometric switch", *IEEE Photonics Technology Letters*, Vol. 16, No. 10, 2350-2352, 2004.

[28] Hong, W; Huang, D; Zhu, G. "Switching window of an SOA-loop-mirror with SOA sped-up by a CW assist light at transparency wavelength", *Optics Communications*, Vol. 238. No.1-3, 151-156, 2004.

[29] Ji, W; Zhang, M; Ye, P. "Simulation of an all-optical XOR gate with a semiconductor optical amplifier Mach-Zehnder interferometer sped up by a continuous-wave assistant light", *Journal of Optical Networking*, Vol. 4, No. 8, 524-530, 2005.

[30] Nielsen, M; Mørk, J. "Bandwidth enhancement of SOA-based switches using optical filtering: theory and experimental verification", *Optics Express*, Vol. 14, No. 3, 1260-1265, 2006

[31] Randel, S; de Melo, AM; Petermann, K; Marembert, V; Schubert, C. "Novel scheme for ultrafast all-optical XOR operation", *Journal of Lightwave Technology*, Vol. 22, No. 12, 2808-2815, 2004.

[32] Sun, H; Wang, Q; Dong, H; Chen, Z; Dutta, NK; Jaques, J; Piccirilli, A. B. "All-optical logic XOR gate at 80 gb/s using SOA-MZI-DI", *IEEE Journal of Quantum Electronics*, Vol. 42, No. 8, 747-751, 2006.

[33] Yang, X; Manning, RJ; Webb, RP. "All-optical 85 Gb/s XOR using dual ultrafast nonlinear interferometers and turbo-switch configuration", *Proceedings European Conference on Optical Communications (ECOC)*, Th. 1.4.2, 2006.

[34] Gutiérrez-Castrejón, R. "Turbo-switched Mach-Zehnder interferometer performance as all-optical signal processing element at 160 Gb/s", *Optics Communications*, Vol. 282, No. 22, 4345-4352, 2009.

[35] Wong, CS; Tsang, HK. "Reduction of bit-patten dependent errors from a semiconductor optical amplifier using an optical delay interferometer", *Optics Communications*, Vol. 232, No. 1-6, 245-249, 2004.

[36] Vazquez, JM; Liu, Z; Li, Y; Tangdiongga, E; Zhang, S; Lenstra, D; Khoe, GD; Dorren, HJS. "Optimization of optical band-pass filters for all-optical wavelength conversion using genetic algorithms", *IEEE Journal of Quantum Electronics*, Vol. 43, No. 1, 57-64, 2007.

[37] Wang, J; Marculescu, A; Li, J; Vorreau, P; Tzadok, S; Ezra, SB; Tsadka, S; Freude, W; Leuthold, J. "Pattern effect removal technique for semiconductor-optical-amplifier-based wavelength conversion", *IEEE Photonics Technology Letters*, Vol. 19, No. 24, 1955-1957, 2007.

[38] Houbavlis, T; Zoiros, KE; Kalyvas, M; Theophilopoulos, G; Bintjas, C; Yiannopoulos, K; Pleros, N; Vlachos, K; Avramopoulos, H; Schares, L; Occhi, L; Guekos, G; Taylor, JR; Hansmann, S; Miller, W. "All-optical signal processing and applications within the Esprit project DO_ALL", *Journal of Lightwave Technology*, Vol. 23, No. 2, 781-801, 2005.
[39] Webb, RP; Manning, RJ; Maxwell, GD; Poustie, AJ. "40 Gbit/s all-optical XOR gate based on hybrid-integrated Mach-Zehnder interferometer", *Electronics Letters*, Vol. 39, No. 1, 79-81, 2003.
[40] Wang, Q; Zhu, G; Chen, H; Jaques, J; Leuthold, J; Piccirilli, AB; Dutta, NK. "Study of all-optical XOR using Mach-Zehnder Interferometer and differential scheme", *IEEE Journal of Quantum Electronics*, Vol. 40, No. 6, 703-710, 2004.
[41] Zhou, Y; Wu, J; Lin, J. "Novel ultrafast all-optical XOR scheme based on Sagnac interferometric structure", *IEEE Journal of Quantum Electronics*, Vol. 41, No. 6, 823-827, 2005.
[42] Berg, TW; Mørk, J. "Saturation and noise properties of quantum-dot optical amplifiers", *IEEE Journal of Quantum Electronics*, Vol. 40, No. 11, 1527-1539, 2004.
[43] Akiyama, T; Ekawa, M; Sugawara, M; Kawaguchi, K; Sudo, H; Kuramata, A; Ebe, H; Arakawa, Y. "An ultrawide-band semiconductor optical amplifier having an extremely high penalty-free output power of 23 dBm achieved with quantum dots", *IEEE Photonics Technology Letters*, Vol. 17, No. 8, 1614-1616, 2005.
[44] Yasuoka, N; Kawaguchi, K; Ebe, H; Akiyama, T; Ekawa, M; Morito, K; Sugawara, M; Arakawa, Y. "Quantum-dot semiconductor optical amplifiers with polarization-independent gains in 1.5-μm wavelength bands", *IEEE Photonics Technology Letters*, Vol. 20, No. 23, 1908-1910, 2008.
[45] Zilkie, AJ; Meier, J; Mojahedi, M; Poole, PJ; Barrios, P; Poitras, D; Rotter, TJ; Yang, C; Stintz, A; Malloy, KJ; Smith, PWE. Aitchison, JS. "Carrier dynamics of quantum-dot, quantum-dash and quantum-well semiconductor optical amplifiers operating at 1.55 μm", *IEEE Journal of Quantum Electronics*, Vol. 43, No. 11, 982-991.
[46] Sugawara, M; Akiyama, T; Hatori, N; Nakata, Y; Ebe, H; Ishikawa, H. "Quantum-dot semiconductor optical amplifiers for high-bit-rate signal processing up to 160 Gbs-1 and a new scheme of 3R regenerators", *Measurement and Science Technology*, Vol. 13, No. 11, 1683-1691, 2002.

[47] Uskov, AV; O'Reilly, EP; Manning, RJ; Webb, RP; Cotter, D; Laemmlin, M; Ledentsov, NN; Bimberg, D. "On ultrafast optical switching based on quantum-dot semiconductor optical amplifiers in nonlinear interferometers", *IEEE Photonics Technology Letters*, Vol. 16, No. 5, 1265-1267, 2004.

[48] Glesk, I; Runser, RJ; Prucnal, PR. "New generation of devices for all-optical communications", *Acta Physica Slovaca*, Vol. 51, No. 2, 151-162, 2001.

[49] Schreieck, RP; Kwakernaak, MH; Jäckel, H; Melchior, H. "All-optical switching at multi-100-Gb/s data rates with Mach-Zehnder interferometer switches", *IEEE Journal of Quantum Electronics*, Vol. 38, No. 8, 1053-1061, 2002.

[50] Sun, H; Wang, Q; Dong, H; Dutta, NK. "All-optical logic performance of quantum-dot semiconductor amplifier-based devices", *Microwave and Optical Technology Letters*, Vol. 48, No. 1, 29-35, 2006.

[51] Han, H; Zhang, M; Ye, P; Zhang, F. "Parameter design and performance analysis of an ultrafast all-optical XOR gate based on quantum dot semiconductor optical amplifiers in nonlinear Mach-Zehnder interferometer", *Optics Communications*, Vol. 281, No. 20, 5140-5145, 2008.

[52] Ben Ezra, Y; Lembrikov, BI; Haridim, M. "Ultrafast all-optical processor based on quantum-dot semiconductor optical amplifiers", *IEEE Journal of Quantum Electronics*, Vol. 45, No. 1, 34-41, 2009.

[53] Rostami, A; Nejad, HBA; Qartavol, RM; Saghai, HR. "Tb/s optical logic gates based on quantum-dot semiconductor optical amplifiers", *IEEE Journal of Quantum Electronics*, Vol. 46, No. 3, 354-360, 2010.

[54] Qasaimeh, O. "An analytical model for quantum dot semiconductor optical amplifiers", *Optics Communications*, Vol. 222, No. 1-6, 277-287, 2003.

[55] Barnham, K; Vvdensky, DD. *Low-dimensional semiconductor structures: Fundamentals and device applications*, Cambridge University Press, 2001.

[56] Scheel, HJ; Capper, P. *Crystal growth technology: From fundamentals and simulation to large-scale production*, Wiley-VCH, Weinheim, 2008.

[57] Ben-Ezra, Y; Lembrikov, BI; Haridim, M. "Acceleration of gain recovery and dynamics of electrons in QD-SOA", *IEEE Journal of Quantum Electronics*, Vol. 41, No. 10, 1268-1273, 2005.

[58] Ben-Ezra, Y; Lembrikov, BI; Haridim, M. "Specific features of XGM in QD-SOA", *IEEE Journal of Quantum Electronics*, Vol. 43, No. 8, 730-737, 2007.

[59] Majer, N; Lüdge, K; Schöll, E. "Cascading enables ultrafast gain recovery dynamics of quantum dot semiconductor optical amplifiers", *Physical Review B*, Vol. 82, art. No. 235301, 2010.

[60] Hamié, A; Hamze, M; Wei, JL; Sharaiha, A; Tang, JM. "Theoretical investigations of quantum-dot semiconductor optical amplifier enabled intensity modulation of adaptively modulated optical OFDM signals in IMDD PON systems", *Optics Express*, Vol. 19, No. 25, 25696-25711, 2011.

[61] Li, X; Li, G. "Comments on "Theoretical analysis of gain recovery time and chirp in QD-SOA" ", *IEEE Photonics Technology Letters*, Vol. 18, No. 22, 2434-2435, 2006.

[62] Uskov, AV; Mørk, J; Tromborg, B; Berg, TW; Magnusdottir, I; O'Reilly, EP. "On high-speed cross-gain modulation without pattern effects in quantum dot semiconductor optical amplifiers", *Optics Communications*, Vol. 227, No. 4-6, 363-369, 2003.

[63] Wong, WM; Blow, KJ. "Travelling-wave model of semiconductor optical amplifier based non-linear loop mirror", *Optics Communications*, Vol. 215, No. 1–3, 169–184, 2003.

[64] Sygletos, S; Spyropoulou, M; Vorreau, P; Bonk, R; Tomkos, I; Freude W; Leuthold, J. "Multi-wavelength regenerative amplification based on quantum-dot semiconductor optical amplifiers", *Proceedings International Conference Transparent Optical Networks (ICTON)*, We.D2.5, 234-237, 2007.

[65] Chattopadhyay, T. "All-optical modified Fredkin gate", *IEEE Journal of Selected Topics in Quantum Electronics*, Vol. 18, No. 2, 585-592, 2012.

[66] Chattopadhyay, T. "All-optical programmable Boolean logic unit using semiconductor optical amplifiers on the Mach-Zehnder interferometer arms switch", *IET Optoelectronics*, Vol. 5, No. 6, 270–280, 2011.

[67] Zilkie, AJ; Meier, J; Mojahedi, M; Helmy, AS; Poole, PJ; Barrios, P; Poitras, D; Rotter, TJ; Yang, C; Stintz, A; Malloy, KJ; Smith PWE; Aitchison, JS. "Time-resolved linewidth enhancement factors in quantum dot and higher-dimensional semiconductor amplifiers operating at 1.55 μm", *Journal of Lightwave Technology*, Vol. 26, No. 11, 1498-1509, 2008.

[68] Nielsen, ML; Mørk, J; Suzuki, R; Sakaguchi, J; Ueno, Y. "Theoretical and experimental study of fundamental differences in the noise

suppression of high-speed SOA-based all-optical switches", *Optics Express*, Vol. 13, No. 13, 5080-5086, 2005.

[69] de Melo, AM; Petermann, K. "On the amplified spontaneous emission noise modeling of semiconductor optical amplifiers", *Optics Communications*, Vol. 281, No. 18, 4598-4605, 2008.

[70] Xao, J-L; Huang, Y-Z. "Numerical analysis of gain saturation, noise figure, and carrier distribution for quantum-dot semiconductor-optical amplifiers", *IEEE Journal of Quantum Electronics*, Vol. 44, No. 5, 448-455, 2008.

[71] Won, YY; Han, S-K. "Amplified spontaneous emission noises in semiconductor optical amplifier wavelength converters using cross-phase modulation", *Microwave and Optical Technology Letters*, Vol. 25, No. 1, 76-81, 2000.

[72] Kotb, A; Ma, S; Chen, Z; Dutta, NK; Said, G. "Effect of amplified spontaneous emission on semiconductor optical amplified based all-optical logic", *Optics Communications*, Vol. 284, No. 24, 5798-5803, 2011.

[73] Bonk, R; Vallaitis, T; Guetlein, J; Meuer, C; Schmeckebier, H; Bimberg, D; Koos, C; Freude, W; Leuthold, J. "The input power dynamic range of a semiconductor optical amplifier and its relevance for access network applications", *IEEE Photonics Journal*, Vol. 3, No. 6, 1039-1053, 2011.

[74] Contestabile, G; Maruta, A; Kitayama, K. "Gain dynamics in quantum-dot semiconductor optical amplifiers at 1550 nm", *IEEE Photonics Technology Letters*, Vol. 22, No. 13, 987-989, 2010.

[75] Maram, RQ; Baghban, H; Rasooli, HS; Ghorbani, R; Rostami, A. "Equivalent circuit model of quantum dot semiconductor optical amplifiers: Dynamic behavior and saturation properties", *Journal of Optics A: Pure and Applied Optics*, Vol. 11, art. no. 105205, 2009.

[76] Berg, TW; Bischoff, S; Magnusdottir, I; Mørk, J. "Ultrafast gain recovery and modulation limitations in self-assembled quantum-dot devices", *IEEE Photonics Technology Letters*, Vol. 13, No. 6, 541-543, 2001.

[77] Gavioli, G; Thomsen, BC; Mikhailov, V; Bayvel, P. "Cascadability properties of optical 3R regenerators based on SOAs", *Journal of Lightwave Technology*, Vol. 25, No. 9, 2766-2775, 2007.

[78] Zhou, D; Kang, K. II; Glesk, I; Prucnal, PR. "An analysis of signal-to-noise ratio and design parameters of a Terahertz Optical Asymmetric Demultiplexer", IEEE *Journal of Lightwave Technology*, Vol. 17, No. 2, 298-307, 1999.

[79] Connelly, MJ. *Semiconductor Optical Amplifiers*, Kluwer, 2002.

[80] Dorren, HJS; Lenstra, D; Liu, Y; Hill, MT; Khoe, GD. "Nonlinear polarization rotation in semiconductor optical amplifiers: Theory and application to all-optical flip-flop memories", *IEEE J. Quantum Electronics*, Vol. 39, No. 1, 141–147, 2003.

[81] Pilai, BSG; Premaratne, M; Abramson, D; Lee, KL; Nirmalathas, A; Lim, C; Shinada, S; Wada, N; Miyazaki, T. "Analytical characterization of optical pulse propagation in polarization-sensitive semiconductor optical amplifiers", *IEEE Journal of Quantum Electronics*, Vol. 42, No. 10, 1062-1077, 2006.

[82] Yasuoka, N; Ebe, H; Kawaguchi, K; Ekawa, M; Sekiguchi, S; Morito, K; Wada, O; Sugawara, M; Arakawa, Y. "Polarization-insensitive quantum dot semiconductor optical amplifiers using strain-controlled columnar quantum dots", *Journal of Lightwave Technology*, Vol. 30, No. 1, 68-75, 2012.

[83] Zhou, E; Zhang, X; Huang, D. "Analysis on dynamic characteristics of semiconductor optical amplifiers with certain facet reflection based on detailed wideband model", *Optics Express*, Vol. 15, No. 14, 9096-9106, 2007.

[84] Sun, J. "Relaxation of facet reflection restrictions in XGM wavelength converters", *Optics Communications*, Vol. 206, No. 1-3, 67-75, 2002.

[85] Huang, A; Whitaker, N; Avramopoulos, H; French, P; Houh, H; Chuang, I. "Sagnac fiber logic gates and their possible applications: a system perspective," *Applied Optics*, Vol. 33, No. 26, 6254–6267, 1994.

[86] Yongpeng, Z; Peida, Y. "Analysis of switching performance of the semiconductor optical amplifier-Mach-Zehnder interferometer based all-optical add/drop multiplexer", *Microwave and Optical Technology Letters*, Vol. 31, No. 2, 77-81, 2001.

[87] Hinton, K; Raskutti, G; Farrell, PM; Tucker, RS. "Switching energy and device size limits on digital photonic signal processing technologies", *IEEE Journal of Selected Topics in Quantum Electronics*, Vol. 14, No. 3, 938-945, 2008.

[88] Ueno, Y; Sakaguchi, J; Nakamoto, R; Nishida, T. "Ultrafast, low-energy consumption, semiconductor-based, all-optical devices", *Proceedings Asia-Pacific Microwave Photonics Conference (APMP)*, 2009.

[89] Pleros, N; Zakynthinos, P; Poustie, A; Tsiokos, D; Bakopoulos, P; Petrantonakis, D; Kanellos, GT; Maxwell, G; Avramopoulos, H. "Optical signal processing using integrated multi-element SOA-MZI

switch arrays for packet switching", *IET Optoelectronics*, Vol. 1, No. 3, 120-126, 2007.

[90] Meuer, C; Langhorst, CS; Bonk, R; Schmeckebier, H; Arsenijevic, D; Fiol, G; Galperin, A,, Leuthold, J; Schubert, C; Bimberg, D. "80 Gb/s wavelength conversion using a quantum-dot semiconductor optical amplifier and optical filtering", *Optics Express*, Vol. 19, No. 6, 5134-5142, 2011.

[91] Tan, H; Mi, Z. Bhattacharya, P; Klotzkin, D. "Dependence of linewidth enhancement factor on duty cycle in InGaAs-GaAs quantum-dot lasers", *IEEE Photonics Technology Letters*, Vol. 20, No. 8, 593-595, 2008.

[92] Kim, SH; Kim, JH; Choi, JW; Son, CW; Byun, YT; Jhon, YM; Lee, S; Woo, DH; Kim, SH. "All-optical half adder using cross gain modulation in semiconductor optical amplifiers", *Optics Express*, Vol. 14, No. 22, 10693-10698, 2006.

[93] Liu, Y; Tangdiongga, E; Li, Z; Zhang, S; de Waardt, H; Khoe, GD; Dorren, HJS. "Error-free all-optical wavelength conversion at 160 Gb/s using a semiconductor optical amplifier and an optical bandpass filter", *Journal of Lightwave Technology*, Vol. 24, No. 1, 230-236, 2006.

[94] Contestabile, G; Maruta, A; Sekiguchi, S; Morito, K; Suqawara, M; Kitayama, K. "Cross-gain modulation in quantum-dot SOA at 1550 nm", *IEEE Journal of Quantum Electronics*, Vol. 46, No. 12, 1696-1703, 2010.

[95] Vázquez, JM; Li, Z; Liu, Y; Tangdiongga, E; Zhang, S; Lenstra, D; Khoe, GD; Dorren, HJS. "Optimization of optical band-pass filters for all-optical wavelength conversion using genetic algorithms", *IEEE Journal of Quantum Electronics*, Vol. 43, No. 1, 57-64, 2007.

[96] Dong, J; Zhang, X; Fu, S; Xu, J; Shum, P. Huang, D. "Ultrafast all-optical signal processing based on single semiconductor optical amplifier and optical filtering", *IEEE Journal of Selected Topics in Quantum Electronics*, Vol. 14, No. 3, 770-778, 2008.

[97] Pauer, M; Winzer, PJ; Leeb, WR. "Bit error probability reduction in direct detection optical receivers using RZ coding", *Journal of Lightwave Technology*, Vol. 19, No. 9, 1255-1262, 2001.

[98] Sun, J; Wang, J. "Simulation of optical NOT gate switching by sum-frequency generation in $LiNbO_3$ waveguides", *Optics Communications*, Vol. 267, No. 1, 187-192, 2006.

[99] Velanas, P; Bogris, A; Syvridis, D. "Operation properties of a reconfigurable photonic logic gate based on cross phase modulation in

highly nonlinear fibers", *Optical Fiber Technology*, Vol. 15, No. 1, 65-73, 2009.

[100] Ta'eed, VG; Eggleton, B. "Ultrafast all-optical signal processing using chalcogenide glass based photonic chips", *Proceedings International Conference Transparent Optical Networks (ICTON)*, WeC1.3, 228-231, 2007.

[101] Dimitriadou, E; Zoiros, KE. "On the feasibility of 320 Gb/s all-optical AND gate using quantum-dot semiconductor optical amplifier-based Mach-Zehnder interferometer", *Progress in Electromagnetics Research B*, Vol. 50, 113-140, 2013.

[102] Dimitriadou, E; Zoiros, KE. "On the feasibility of ultrafast all-optical NAND gate using single quantum-dot semiconductor optical amplifier-based Mach-Zehnder interferometer", *Optics & Laser Technology*, Vol. 44, No. 6, 1971–1981, 2012.

[103] Dimitriadou, E; Zoiros, KE. "Proposal for all-optical NOR gate using single quantum-dot semiconductor optical amplifier-based Mach-Zehnder interferometer", *Optics Communications*, Vol. 285, No. 7, 1710-1716, 2012.

[104] Mano, MM; Ciletti, MD. *Digital design*, 4th Ed; Prentice Hall, 2006.

[105] Porzi, C; Scaffardi, M; Poti, L; Bogoni, A. "Optical digital signal processing in a single SOA without assist probe light", *IEEE Journal of Selected Topics in Quantum Electronics*, Vol. 16, No. 5, 1469–1475, 2010.

[106] Leuthold, J; Wang, J; Vallaitis, T; Koos, C; Bonk, R; Marculescu, A; Vorreau, P; Sygletos, S; Freude, W. "New approaches to perform all-optical signal regeneration", *Proceedings International Conference Transparent Optical Networks (ICTON)*, Vol. 2, 222–225, 2007.

[107] Yang, W; Zhang, M; Ye, P. "Analysis of all-optical demultiplexing from 160/320 Gbit/s to 40 Gbit/s using quantum-dot semiconductor optical amplifiers assisted Mach-Zehnder interferometer", *Microwave Optical Technology Letters*, Vol. 52, No. 7, 1629-1633, 2010.

[108] O'Driscoll, I; Piwonski, T; Schleussner, CF; Houlihan, J; Huyet, G; Manning, RJ. "Electron and hole dynamics of InAs/GaAs quantum dot semiconductor optical amplifiers", *Applied Physics Letters*, Vol. 91, No. **7,** art. no. 071111/ 1–3, 2007.

[109] Borri, P; Langbein, W; Schneider, S; Woggon, U; Sellin, RL; Ouyang, D; Bimberg, D. "Exciton relaxation and dephasing in quantum-dot amplifiers from room to cryogenic temperature", *IEEE Journal of Selected Topics in Quantum Electronics*, Vol. 8, No. 5, 984–991, 2002.

[110] Agrawal, GP. *Fiber-Optic Communication Systems*, 3rd ed, Wiley, 2002.

[111] Vardakas, JS; Zoiros, KE. "Performance investigation of all-optical clock recovery circuit based on Fabry-Pérot filter and SOA-assisted Sagnac switch", *Optical Engineering*, Vol. 46, No.8, art. no. 085005/ 1-21, 2007.

[112] Gutiérrez-Castrejón, R; Occhi, L; Schares, L; Guekos, G. "Recovery dynamics of cross-modulated beam phase in semiconductor amplifiers and applications to all-optical signal processing", *Optics Communications*, Vol. 195, No. 1-4, 167-177, 2001.

[113] MacWilliams, FJ; Sloane, NJA. "Pseudo-random sequences and arrays", *Proceedings IEEE*, Vol. 64, No. 12, 1715–1729, 1976.

[114] Gayen, DK; Bhattachryya, A; Roy, JN. "Ultrafast all-optical half adder using quantum-dot semiconductor optical amplifier-based Mach-Zehnder interferometer", *Journal of Lightwave Technology*, Vol. 30, No. 21, 3387-3393, 2012.

[115] Ma, S; Sun, H; Chen, Z; Dutta, NK. "High speed all-optical PRBS generation based on quantum-dot semiconductor optical amplifiers", *Opt. Express*, Vol. 17, No. 21, 18469-18477, Sep. 2009.

[116] Li, W; Ma, S; Hu, H; Dutta, NK. "All optical latches using quantum-dot semiconductor optical amplifier", *Optics Communications*, Vol. 285, 5138-5143, 2012.

ISBN: 978-1-63483-023-2

BIBLIOGRAPHY

An introduction to semi-tensor product of matrices and its applications LCCN: 2012554769 Type of material: Book Personal name: Cheng, Dai-Zhan. Main title: An introduction to semi-tensor product of matrices and its applications / Daizhan Cheng, Hongsheng Qi, Yin Zhao. Published/Created: Singapore; Hackensack, N.J.: World Scientific, c2012. Description: xxii, 587 p.: ill.; 24 cm. ISBN: 9789814374682 (cloth) 9814374687 (cloth) LC classification: QA329 .C476 2012 Related names: Qi, Hongsheng, Ph.D. Zhao, Yin. Contents: 1. Multi-dimensional data -- 2. Semi-tensor product of matrices -- 3. Multilinear mappings among vector spaces -- 4. Right and general semi-tensor products -- 5. Rank, pseudo-inverse, and positivity of STP -- 6. Matrix expression of logic -- 7. Mix-valued logic -- 8. Logical matrix, fuzzy set and fuzzy logic -- 9. Fuzzy relational equation -- 10. Fuzzy control with coupled fuzzy relations -- 11. Representation of boolean functions -- 12. Decomposition of logical functions -- 13. Boolean calculus -- 14. Lattice, graph, and universal algebra -- 15. Boolean network -- 16. Boolean control system -- 17. Game theory -- 18. Multi-variable polynomials -- 19. Some applications to differential geometry and algebra -- 20. Morgan's problem -- 21. Linearization of nonlinear control systems -- 22. Stability region of dynamic systems -- Appendix A. Numerical algorithms. Subjects: Operator theory. Matrices. Spectral theory (Mathematics) Notes: Includes bibliographical references (p.

573-581) and index. Dewey class no.: 515.724 Other system no.: (OCoLC)ocn805312364 CALL NUMBER: QA329 .C476 2012 Copy 1

Analysis of boolean functions LCCN: 2013050033 Type of material: Book Personal name: O'Donnell, Ryan, 1979- author. Main title: Analysis of boolean functions / Ryan O'Donnell, Carnegie Mellon University, Pittsburgh, Pennsylvania. Published/Produced: New York, NY: Cambridge University Press, 2014. Projected pub date: 1403 Description: pages cm ISBN: 9781107038325 (hardback: acid-free paper) LC classification: QA76.9.M35 O36 2014 Summary: "Boolean functions are perhaps the most basic objects of study in theoretical computer science. They also arise in other areas of mathematics, including combinatorics, statistical physics, and mathematical social choice. The field of analysis of Boolean functions seeks to understand them via their Fourier transform and other analytic methods. This text gives a thorough overview of the field, beginning with the most basic definitions and proceeding to advanced topics such as hypercontractivity and isoperimetry. Each chapter includes a "highlight application" such as Arrow's theorem from economics, the Goldreich-Levin algorithm from cryptography/learning theory, Hastad's NP-hardness of approximation results, and "sharp threshold" theorems for random graph properties. The book includes roughly 450 exercises and can be used as the basis of a one-semester graduate course. It should appeal to advanced undergraduates, graduate students, and researchers in computer science theory and related mathematical fields"-- Provided by publisher. Subjects: Computer science--Mathematics. Algebra, Boolean. COMPUTERS / General. Notes: Includes bibliographical references and index. Dewey class no.: 004.01/51 Other class no.: COM000000 Content type: text Media type: unmediated Carrier type: volume Library of Congress Holdings Information not available.

Boolean functions in coding theory and cryptography LCCN: 2011035308 Type of material: Book Personal name: Logachev, Oleg A. Uniform title: Bulevy funktsii v teorii kodirovaniia i kriptologii. English Main title: Boolean functions in coding theory and cryptography / O.A. Logachev, A.A. Sal'nikov, V.V. Yashchenko; translated by Svetla Nikova. Published/Created: Providence, R.I.: American Mathematical Society, c2012. Description: xiv, 334 p.; 26 cm. ISBN: 9780821846803 (alk. paper) LC classification: QA268 .L6413 2011 Related names: Sanikov, A. A. (Alekseí Aleksandrovich) IAshchenko, V. V. Subjects: Coding theory. Cryptography. Algebra, Boolean. Information and communication, circuits -- Research exposition (monographs, survey articles). Information and communication, circuits -- Communication, information -- Cryptography. Information and communication, circuits -- Circuits, networks -- Switching theory, application of Boolean algebra; Boolean functions. Notes: Includes bibliographical references and index. Series: Translations of mathematical monographs; v. 241 Dewey class no.: 003/.54 Language code: eng rus Other class no.: 94-02 94A60 94C10 CALL NUMBER: QA268 .L6413 2011 Copy 1 Request in: Jefferson or Adams Building Reading Rooms CALL NUMBER: QA268 .L6413 2011 Copy 2

Boolean functions: theory, algorithms, and applications LCCN: 2011009690 Type of material: Book Personal name: Crama, Yves, 1958- Main title: Boolean functions: theory, algorithms, and applications / Yves Crama, Peter L. Hammer. Published/Created: Cambridge; New York: Cambridge University Press, 2011. Description: xxi, 687 p.: ill.; 24 cm. Links: Cover image http://assets.cambridge.org/97805218/47513/cover/9780521847513.jpg ISBN: 0521847516 (hardback) 9780521847513 (hardback) LC classification: QA341 .C73 2011 Related names: Hammer, P. L., 1936-2006. Contents: Fundamental concepts and applications -- Boolean equations -- Prime implicants and minimal DNFs / Peter L. Hammer and Alexander Kogan -- Duality theory / Yves Crama and Kazuhisa Makino -- Quadratic functions / Bruno Simeone -- Horn functions / Endre Boros -- Orthogonal forms and shellability -- Regular functions -- Threshold functions -- Red-once functions / Martin C. Golumbic and Vladimir Gurvich -- Characterizations of special classes by functional equations / Lisa Hellerstein -- Partially defined Boolean functions / Toshihide Ibaraki --

Pseudo-Boolean functions -- Graphs and hypergraphs -- Algorithmic complexity -- JBool: a software tool / Claude Benzaken and Nadia Brauner. Subjects: Algebraic functions. Algebra, Boolean. Notes: "Written by prominent experts in the field, this monograph provides the first comprehensive and unified presentation of the structural, algorithmic, and applied aspects of the theory of Boolean functions. The book focuses on algebraic representations of Boolean functions, especially disjunctive and conjunctive normal form representations. It presents with in this framework the fundamental elements of the theory (Boolean equations and satisfiability problems, prime implicants and associated short representations, dualization), an indepth study of special classes of Boolean functions (quadratic, Horn, shellable, regular, threshold, read-once functions and their characterization by functional equations), and two fruitful generalizations of the concept of Boolean functions (partially defined functions and pseudo-Boolean functions). Several topics are presented here in book form forthe first time. Because of the unique depth and breadth of the unified treatment that it provides and its emphasis on algorithms and applications, this monograph will have special appeal for researchers and graduate students in discrete mathematics, operations research, computer science, engineering, and economics"--Provided by publisher. Includes bibliographical references and index. Series: Encyclopedia of mathematics and its applications; 142 Encyclopedia of mathematics and its applications; v. 142. Dewey class no.: 511.3/24 Other class no.: MAT017000 Other system no.: (OCoLC)ocn711777210 CALL NUMBER: QA341 .C73 2011 Copy 1 Request in: Jefferson or Adams Building Reading Rooms CALL NUMBER: QA341 .C73 2011 Copy 2

Boolean models and methods in mathematics, computer science, and engineering LCCN: 2010017816 Type of material: Book Main title: Boolean models and methods in mathematics, computer science, and engineering / edited by Yves Crama, Peter L. Hammer. Published/Created: Cambridge; New York: Cambridge University Press, 2010. Description: xviii, 759 p.: ill.; 25 cm. Links: Cover image http://assets.cambridge.org/97805218/47520/cover/9780521847520.jpg Table of contents only http://www.loc.gov/catdir/enhancements/fy1009/2010017816-t.html Publisher description http://www.loc.gov/catdir/enhancements/fy1009/2010017816-d.html Contributor biographical information http://www.loc.gov/catdir/enhancements/fy1009/2010017816-b.html ISBN: 0521847524 9780521847520 LC classification: QA10.3 .B658 2010 Related names: Crama, Yves, 1958- Hammer, P. L., 1936-2006. Summary: "This collection of papers presents a series of in-depth examinations of a variety of advanced topics related to Boolean functions and expressions. The chapters are written by some of the most prominent experts in their respective fields and cover topics ranging from algebra and propositional logic to learning theory, cryptography, computational complexity, electrical engineering, and reliability theory. Beyond the diversity of the questions raised and investigated in different chapters, a remarkable feature of the collection is the common thread created by the fundamental language, concepts, models, and tools provided by Boolean theory. Many readers will be surprised to discover the countless links between seemingly remote topics discussed in various chapters of the book. This text will help them draw on such connections to further their understanding of their own scientific discipline and to explore new avenues for research"--Provided by publisher. "This collection of papers presents a series of in-depth examinations of a variety of advanced topics related to Boolean functions and expressions. The chapters are written by some of the most prominent experts in their respective fields and cover topics ranging from algebra and propositional logic to learning theory, cryptography, computational complexity, electrical engineering, and reliability theory. Beyond the diversity of the questions raised and investigated in different chapters, a remarkable feature of the collection is the common thread created by the fundamental language, concepts, models, and tools provided by Boolean theory"-- Provided by publisher. Contents: Algebraic structures. Compositions and clones of Boolean functions / Reinhard Pöschel and Ivo Rosenberg -- Decomposition of

Boolean functions / Jan C. Bioch -- Logic. Proof theory / Alasdair Urquhart -- Probabilistic analysis of satisfiability algorithms / John Franco -- Optimization methods in logic / John Hooker -- Learning theory and cryptography. Probabilistic learning and Boolean functions / Martin Anthony -- Learning Boolean functions with queries / Robert H. Sloan, Balázs Szörényi, and György Turán -- Boolean functions for cryptography and error-correcting codes / Claude Carlet -- Vectorial Boolean functions for cryptography / Claude Carlet -- Graph representations and efficient computation models. Binary decision diagrams / Beate Bollig ... [et al.] -- Circuit complexity / Matthias Krause and Ingo Wegener -- Fourier transforms and threshold circuit complexity / Jehoshua Bruck -- Neural networks and Boolean functions / Martin Anthony -- Decision lists and related classes of Boolean functions / Martin Anthony -- Applications in engineering. Hardware equivalence and property verification / J.-H. Roland Jiang and Tiziano Villa -- Synthesis of multi-level Boolean networks / Tiziano Villa ... [et al.] -- Boolean aspects of network reliability / Charles J. Colbourn. Subjects: Algebra, Boolean. Probabilities. Notes: Series numbering from jacket. Includes bibliographical references. Series: Encyclopedia of mathematics and its applications; [134] Encyclopedia of mathematics and its applications; v. 134. Dewey class no.: 511.3/24 National bib no.: GBB076675 National bib agency no.: 015584825 Other system no.: (OCoLC)ocn607985575 CALL NUMBER: QA10.3 .B658 2010 Copy 2 Request in: Jefferson or Adams Building Reading Rooms CALL NUMBER: QA10.3 .B658 2010 Copy 1

Cryptology and network security: 11th international conference, CANS 2012, Darmstadt, Germany, December 12-14, 2012: proceedings LCCN: 2012953024 Type of material: Book Meeting name: International Conference on Cryptology and Network Security (11th: 2012: Darmstadt, Germany) Main title: Cryptology and network security: 11th international conference, CANS 2012, Darmstadt, Germany, December 12-14, 2012: proceedings / Josef Pieprzyk, Ahmad-Reza Sadeghi, Mark Manulis (eds.). Published/Produced: Heidelberg

[Germany]: Springer, [2012]. ©2012. Description: xvi, 309 pages: illustrations; 24 cm. ISBN: 9783642354038 (pbk.: alk. paper) 3642354033 (pbk.: alk. paper) LC classification: QA76.9.A25 I5538 2012 Variant title: CANS 2012 Related names: Pieprzyk, Josef, 1949- editor. Sadeghi, Ahmad-Reza, editor. Manulis, Mark, editor. Contents: Cryptanalysis -- Conditional Differential Cryptanalysis of Grain-128a / Michael Lehmann, Willi Meier -- A Real-Time Key Recovery Attack on the Lightweight Stream Cipher A2U2 / Zhenqing Shi, Xiutao Feng, Dengguo Feng, Chuankun Wu -- A Simple Key-Recovery Attack on McOE-X / Florian Mendel, Bart Mennink, Vincent Rijmen, Elmar Tischhauser -- Cryptanalysis of a Lattice-Knapsack Mixed Public Key Cryptosystem / Jun Xu, Lei Hu, Siwei Sun, Ping Wang -- Biclique Cryptanalysis of TWINE / Mustafa Çoban, Ferhat Karakoç, Özkan Boztaş -- Differential and Linear Attacks on the Full WIDEA-n Block Ciphers (under Weak Keys) / Jorge Nakahara Jr. -- Improved Linear Analysis on Block Cipher MULTI2 / Yi Lu, Liping Ding, Yongji Wang -- Fixed Points of Special Type and Cryptanalysis of Full GOST / Orhun Kara, Ferhat Karakoç -- Network Security -- Attacking Animated CAPTCHAs via Character Extraction / Vu Duc Nguyen, Yang-Wai Chow, Willy Susilo -- Analysis of Rogue Anti-Virus Campaigns Using Hidden Structures in k-Partite Graphs / Orestis Tsigkas, Dimitrios Tzovaras -- Mobile Evil Twin Malnets - The Worst of Both Worlds / Christian Szongott, Benjamin Henne, Matthew Smith -- Firm Grip Handshakes: A Tool for Bidirectional Vouching / Omer Berkman, Benny Pinkas, Moti Yung -- Cryptographic Protocols -- Group Key Establishment: Adding Perfect Forward Secrecy at the Cost of One Round / Kashi Neupane, Rainer Steinwandt, Adriana Suárez Corona -- Applicability of OR-Proof Techniques to Hierarchical Identity-Based Identification / Atsushi Fujioka, Taiichi Saito, Keita Xagawa -- LiBrA-CAN: A Lightweight Broadcast Authentication Protocol for Controller Area Networks / Bogdan Groza, Stefan Murvay, Anthony van Herrewege, Ingrid Verbauwhede -- Efficient Verification of Input Consistency in Server-Assisted Secure Function Evaluation / Vladimir Kolesnikov, Ranjit

Kumaresan, Abdullatif Shikfa -- Fast and Private Computation of Cardinality of Set Intersection and Union / Emiliano De Cristofaro, Paolo Gasti, Gene Tsudik -- Encryption -- Fast and Secure Root Finding for Code-Based Cryptosystems / Falko Strenzke -- Strong Privacy for RFID Systems from Plaintext-Aware Encryption / Khaled Ouafi, Serge Vaudenay -- How to Enhance the Security on the Least Significant Bit / Atsuko Miyaji, Yiren Mo -- S-Box Theory -- Improvement in Non-linearity of Carlet-Feng Infinite Class of Boolean Functions / Mansoor Ahmed Khan, Ferruh Özbudak -- Some Representations of the S-Box of Camellia in $GF(((2^2)^2)^2)$ / Alberto F. Martínez-Herrera, J. Carlos Mex-Perera, Juan A. Nolazco-Flores. Subjects: Data encryption (Computer science)--Congresses. Computer networks--Security measures--Congresses. Notes: Includes bibliographical references and index. Series: Lecture notes in computer science, 0302-9743; 7712 LNCS sublibrary. SL 4, Security and cryptology Lecture notes in computer science; 7712. LNCS sublibrary. SL 4, Security and cryptology. Other system no.: (OCoLC)ocn818734651 Invalid ISBN: 9783642354045 (e-ISBN) Content type: text Media type: unmediated Carrier type: volume CALL NUMBER: QA76.9.A25 I5538 2012 Copy 1

Digital logic design: a rigorous approach LCCN: 2012015723 Type of material: Book Personal name: Even, Guy. Main title: Digital logic design: a rigorous approach / Guy Even, Tel Aviv University, Israel, Moti Medina, Tel Aviv University, Israel. Published/Created: Cambridge: Cambridge University Press, 2012. Description: xx, 348 pages: illustrations; 26 cm Links: Contributor biographical information http://www.loc.gov/catdir/enhancements/fy1210/2012015723-b.html Publisher description http://www.loc.gov/catdir/enhancements/fy1210/2012015723-d.html Table of contents only http://www.loc.gov/catdir/enhancements/fy1210/2012015723-t.html ISBN: 9781107027534 (Hardback) LC classification: TK7868.L6 E94 2012 Related names: Medina, Moti, 1979- Summary: "Chapter 1 Sets and Functions This chapter introduces two major notions: sets and functions. We are all familiar with real functions, for example f(x} = 2x + 1 and g(x} = sin(x). Here the approach is somewhat different. The first difference is that we do not limit the discussion to the set of real numbers. Instead, we consider arbitrary sets, and are mostly interested in sets that contain only a finite number of elements. The second difference is that we do not define a 'rule" for assigning a value for each x. Instead, a function is simply a list of pairs (x,y), where y denotes the value of the function when the argument equals x. The definition of functions relies on the definitions of sets and relations over sets. That is why we need to define various operations over sets such as: union, intersection, complement, and Cartesian product. The focus of this book is Boolean functions. Boolean functions are a special family of functions. Their arguments and values are finite sequences of zero and ones (also called bits). In this chapter we show how to represent a Boolean function by a truth table and multiplication tables. Other representations presented later in the book are: Boolean formulas and combinational circuits"-- Provided by publisher. Subjects: Logic design. COMPUTERS / General. Notes: Includes bibliographical references (page 343) and index. Dewey class no.: 621.39/5 Other class no.: COM000000 Content type: text Media type: unmediated Carrier type: volume CALL NUMBER: TK7868.L6 E94 2012 Copy 1 Request in: Jefferson or Adams Building Reading Rooms CALL NUMBER: TK7868.L6 E94 2012 Copy 2

Memory-based logic synthesis LCCN: 2011922264 Type of material: Book Personal name: Sasao, Tsutomu, 1950- author. Main title: Memory-based logic synthesis / Tsutomu Sasao. Published/Produced: New York: Springer, [2011] ©2011 Description: xii, 189 pages: illustrations; 24 cm Links: Publisher description http://www.loc.gov/catdir/enhancements/fy1306/2011922264-d.html Table of contents only http://www.loc.gov/catdir/enhancements/fy1306/2011922264-t.html ISBN: 9781441981035 1441981039 LC classification: TK7868.L6 S264 2011 Summary: "This book describes the synthesis of logic functions using memories. It is useful to design field programmable gate arrays (FPGAs) that contain both small-scale memories, called look-up tables (LUTs), and medium-scale memories, called embedded memories. This is a valuable reference for both FPGA system designers and CAD tool developers, concerned with logic synthesis for FPGAs. Anyone using logic gates to design logic circuits, you can benefit from the methods described in this book. Describes in detail the synthesis of logic functions using memories; Introduces a look-up tables (LUT) cascade as a new architecture for logic synthesis; Shows logic design methods for index generation functions; Introduces C-measure, which specifies the complexity of Boolean functions; Introduces hash-based design methods, which efficiently synthesize index generation functions by pairs of smaller memories and can be applied to IP address tables, packet filtering, terminal access controllers, memory patch circuits, virus scanning circuits, intrusion detection circuits, fault map of memories, code converters and pattern matching" --Publisher. Contents: Introduction -- Basic elements -- Definitions and basic properties -- MUX-based synthesis -- Cascade-based synthesis -- Encoding method -- Functions with small C-measures -- C-measure of sparse functions -- Index generation functions -- Hash-based synthesis -- Reduction of the number of variables -- Various realizations -- Conclusions. Subjects: Logic circuits. Field programmable gate arrays--Design and construction. Semiconductor storage devices. Notes: Includes bibliographical references (pages 179-186) and index. Dewey class no.: 621.39 National bib agency no.: 015679265 Other system no.: (OCoLC)ocn690089488 Invalid ISBN: 9781441981042 (e-book) 1441981047 (e-book) Content type: text Media type: unmediated Carrier type: volume CALL NUMBER: TK7868.L6 S264 2011 Copy 1 Request in: Jefferson or Adams Building Reading Rooms CALL NUMBER: TK7868.L6 S264 2011 Copy 2

Noise sensitivity of boolean functions and percolation LCCN: 2014023801 Type of material: Book Personal name: Garban, Christophe, 1982- Main title: Noise sensitivity of boolean functions and percolation / Christophe Garban, ICJ, Université Lyon, Jeffrey E. Steif, Chalmers University of Technology, Gothenberg. Published/Produced: New York, NY: Cambridge University Press, 2015. Description: xvii, 203 pages: illustrations; 24 cm. ISBN: 9781107076433 (hardback) 1107076439 (hardback) 9781107432550 (paperback) 1107432553 (paperback) LC classification: QC174.8 .G36 2015 Related names: Steif, Jeffrey E. Subjects: Statistical physics--Textbooks. Percolation (Statistical physics)--Textbooks. Algebra, Boolean--Textbooks. Notes: Includes bibliographical references (pages 197-202) and index. Series: Institute of Mathematical Statistics textbooks Dewey class no.: 530.13 Content type: text Media type: unmediated Carrier type: volume CALL NUMBER: QC174.8 .G36 2015 Copy 1

Tractability LCCN: 2012276355 Type of material: Book Main title: Tractability / edited by Lucas Bordeaux (Microsoft Research), Youssef Hamadi (Microsoft Research), Pushmeet Kohli (Microsoft Research). Published/Produced: Cambridge: Cambridge University Press, [2014] ©2014 Description: xxi, 377 pages: illustrations; 26 cm ISBN: 9781107025196 hardback 1107025192 LC classification: QA76.9.A43 .T733 2014 Related names: Bordeaux, Lucas, editor of compilation. Hamadi, Youssef (Computer science researcher), editor of compilation. Kohli, Pushmeet, editor of compilation. Summary: An overview of the techniques developed to circumvent computational intractability, a key challenge in many areas of computer science. Contents: Cover; Tractability; Title Page; Copyright Page; Contents; Contributors; Introduction; Part 1: Graphical Structure; 1 Treewidth and Hypertree Width; 1.1 Treewidth; 1.2 Hypertree width; 1.3 Applications of hypertree width; 1.4 Beyond (hyper)tree decompositions; 1.5 Tractability frontiers (for CSPs); 1.6 Conclusion; References; 2 Perfect Graphs and Graphical Modeling; 2.1 Berge Graphs and Perfect Graphs; 2.2 Computational Properties of Perfect Graphs; 2.3 Graphical Models; 2.4 Nand Markov Random Fields; 2.5 Maximum Weight Stable Set; 2.6 Tractable Graphical Models; 2.7 Discussion. 2.8 Acknowledgments; 2.9 Appendix; References; Part 2: Language Restrictions; 3 Submodular Function Maximization; 3.1 Submodular Functions; 3.2 Greedy Maximization of Submodular Functions; 3.3 Beyond the Greedy Algorithm: Handling More Complex Constraints; 3.4 Online Maximization of Submodular Functions; 3.5 Adaptive Submodularity; 3.6 Conclusions; References; 4

Tractable Valued Constraints; 4.1 Introduction; 4.2 Constraint Satisfaction Problems; 4.3 Valued Constraint Satisfaction Problems; 4.4 Examples of Valued Constraint Languages; 4.5 Expressive Power. 4.6 Submodular Functions and Multimorphisms; 4.7 Conservative Valued Constraint Languages; 4.8 A General Algebraic Theory of Complexity; 4.9 Conclusions and Open Problems; References; 5 Tractable Knowledge Representation Formalisms; 5.1 Introduction; 5.2 A Motivating Example; 5.3 Negation Normal Form; 5.4 Structured Decomposability; 5.5 (X, Y)-Decompositions of Boolean Functions; 5.6 Sentential Decision Diagrams; 5.7 The Process of Compilation; 5.8 Knowledge Compilation in Probabilistic Reasoning; 5.9 Conclusion; References; Part 3: Algorithms and their Analysis. 6 Tree-Reweighted Message Passing; 6.1 Introduction; 6.2 Preliminaries; 6.3 Sequential Tree-Reweighted Message Passing (TRW-S); 6.4 Analysis of the Algorithm; 6.5 TRW-S with Monotonic Chains; 6.6 Summary of the TRW-S Algorithm; 6.7 Related Approaches; 6.8 Conclusions and Discussion; References; 7 Tractable Optimization in Machine Learning; 7.1 Introduction; 7.2 Background; 7.3 Smooth Convex Optimization; 7.4 Nonsmooth Convex Optimization; 7.5 Stochastic Optimization; 7.6 Summary; References; 8 Approximation Algorithms; 8.1 Introduction; 8.2 Combinatorial Algorithms. 8.3 Linear Programming Based Algorithms; 8.4 Semi-Definite Programming Based Algorithms; 8.5 Algorithms for Special Instances; 8.6 Metric Embeddings; 8.7 Hardness of Approximation; References; 9 Kernelization Methods for Fixed-Parameter Tractability; 9.1 Introduction; 9.2 Basic Definitions; 9.3 Classical Techniques; 9.4 Recent Upper Bound Machinery; 9.5 Conclusion; References; Part 4: Tractability in Some Specific Areas; 10 Efficient Submodular Function Minimization for Computer Vision; 10.1 Labeling Problems in Computer Vision; 10.2 Markov and Conditional Random Fields; 10.3 Minimizing Energy Functions for MAP Inference. Subjects: Computer algorithms. Combinatorial optimization--Data processing. Computational complexity. Notes: Includes bibliographical references.

Content type: text Media type: unmediated Carrier type: volume Shelf Location: FLM2014 161361 CALL NUMBER: QA76.9.A43 .T733 2014 OVERFLOWA5S

INDEX

A

access, 112, 126
acid, 118
adjustment, 63, 80
algorithm, 71, 118
amplitude, 62, 63, 74, 77, 82, 89, 93, 94, 95, 99, 103, 104
anti-reflection coating, 79
anti-symmetry, vii, 1, 37
Asia, 113
assessment, 95
assets, 119, 121
automation, vii

B

bandwidth, 58, 61, 62, 65, 78, 83
beams, 62
benefits, 90, 99, 105
BI, 110, 111
bias, 63, 80, 93
Boolean functions, vii, 1, 2, 3, 10, 17, 24, 39, 52, 57, 59, 60, 61, 62, 65, 68, 73, 76, 77, 78, 79, 80, 81, 82, 84, 86, 87, 88, 94, 99, 100, 104, 118, 119, 121, 125, 126

C

CAD, 125
calculus, 117
candidates, 61, 65
classes, 119, 122
classification, 105, 117, 118, 119, 121, 123, 124, 125, 127
coding, 82, 84, 114, 118
combinatorics, 118
commercial, 79, 105
communication, 60, 103, 119
compilation, 127
complement, 60, 74, 76, 81, 82, 84, 85, 86, 87, 89, 91, 125
complexity, 59, 61, 78, 83, 90, 120, 121, 126, 128
composition, 72
computation, 122
computer, 69, 71, 118, 120, 124, 127
computing, 58, 105
conduction, 66, 67
conference, 122
configuration, viii, 58, 59, 60, 61, 70, 78, 80, 81, 84, 85, 94, 108
confinement, 63
Congress, 118
conjugate symmetry, vii, 1, 37, 39, 43, 44
conservation, 74
constituents, 73, 89
construction, 62, 93, 105, 126
consumption, 59, 76, 80
copyright, 127
correlation, 63

cost, 59, 62, 83, 90
cryptography, 118, 121

D

data rates, viii, 58, 59, 60, 62, 68, 70, 72, 80, 92, 110
deduction, 86, 87
degradation, 82
demonstrations, 100
depth, 120, 121
derivatives, 68, 71
designers, 125
detection, 114, 126
differential equations, 69, 71
digital information, viii, 57
dispersion, 87
distribution, 72, 112
diversity, 121
dynamic systems, 117

E

economics, 118, 120
electron(s), 66, 69, 92, 93, 103, 110
emission, 61, 63, 68, 76, 112
encryption, 124
energy, 61, 63, 65, 66, 67, 69, 74, 79, 80, 113
energy consumption, 61, 113
engineering, 120
environment, 58
environmental change, 62
equilibrium, 66, 71, 81
erbium, 61, 92
evolution, 77, 105
examinations, 121
execution, viii, 58, 60, 71, 99
exotic symmetry, vii, 1, 2
exploitation, 64, 65, 78, 105
extinction, 77, 83, 86, 93, 94, 95
extraction, 106

F

fabrication, 66, 79
fiber(s), 60, 61, 64, 76, 78, 79, 81, 86, 87, 92, 105, 106, 113
fiber optics, 105
filters, 108, 114
flexibility, 64
fluctuations, 62, 63, 72, 94, 99

G

geometry, 117
Germany, 122
glasses, 86
graduate students, 118, 120
graph, 117, 118
Greece, 57
group theory, vii, 1, 2
growth, 66, 78, 110
guidelines, viii, 58, 76, 96, 104

H

hardness, 118
height, 97
hierarchical symmetry, vii, 1, 3, 24, 25, 34
high speed switching, viii, 58
host, 60
hybrid, 109
hyper-symmetry, vii, 1, 51

I

ideal, 65, 77, 79, 86
image, 84, 119, 120
InAs/GaAs, 115
input signal, 61, 89
insertion, 106
integration, 61, 62, 65, 81
integrity, 82
interference, 77
inversion, 60, 77, 79, 81, 84, 85, 86, 87

IP address, 126
islands, 66
Israel, 124
issues, viii, 58, 68, 76

K

kinetics, 67
Kronecker symmetry, vii, 1, 38

L

lasers, 114
latency, 61
laws, 89
leakage, 79
learning, 118, 121
lifetime, 63, 69
light, vii, viii, 57, 61, 63, 76, 78, 90, 105, 107, 108, 115
light-wave modulated signals, vii
lithium, 60, 61, 81
logic gates, vii, 57, 105, 106, 110, 113, 126

M

Mach-Zehnder interferometer, viii, 57, 59, 65, 104, 106, 107, 108, 109, 110, 111, 113, 115, 116
magnitude, 73, 74, 80
mass, 66
materials, 86, 105
mathematics, vii, 118, 120
matrix, vii, 117
matrix-based symmetry, vii, 1, 51
matter, 89
measurement, 77
measurements, 83, 86, 93
media, 73
memory, 69, 126
Microsoft, 71, 127
models, 120
modules, 65, 80, 105
MR, 105
multiplication, 125
MZI, viii, 57, 59, 65, 68, 71, 72, 73, 75, 77, 78, 79, 80, 81, 84, 87, 88, 89, 90, 91, 92, 93, 94, 95, 96, 98, 99, 100, 105, 108, 113

N

null, 74, 77, 89

O

obstacles, 59
operating system, 71
operations, viii, 58, 72, 79, 86, 104, 106, 120, 125
operations research, 120
optical communications, 105, 110
optical fiber, 58
optical networks, 65
optical properties, 59, 66, 67, 73, 74, 78, 89, 93
optical pulses, 60, 107
optimization, 63, 78, 128

P

Pacific, 113
parity, 100, 101, 102, 103, 104
percolation, 126
permutation-based symmetry, vii, 1
phase shifts, 107
photonics, viii, 57
physical properties, 65
physics, 118, 127
platform, 58, 60, 81, 92
playing, 83
PM, 113
polarity, 84
polarization, 65, 76, 78, 83, 109, 113
population, 68, 77
probability, 68, 114
probe, 77, 81, 82, 84, 87, 92, 93, 115
programming, 71
project, 109

propagation, 63, 65, 67, 68, 71, 78, 81, 113
pumps, 85

Q

QD-SOAs, viii, 57, 59, 65, 66, 67, 71, 72, 73, 74, 76, 78, 79, 80, 92, 93, 95, 99, 103, 105
quantum dot(s), 109, 110, 111, 112, 113, 115
quantum-dot semiconductor optical amplifiers, viii, 57, 59, 105, 110, 111, 112, 115, 116

R

real numbers, 125
reality, 79
reasoning, 78, 87
recombination, 62, 66
recovery, 62, 63, 65, 66, 67, 77, 82, 93, 107, 110, 111, 112, 116
reflectivity, 68, 76, 79
refractive index, 83
regeneration, 62, 65, 115
rejection, 83
relaxation, 66, 67, 69, 92, 93, 103, 115
relevance, 112
reliability, 62, 121
requirements, 69, 77, 80, 82, 96, 97, 98, 99, 103
researchers, 118, 120
resolution, 69
resources, 90
response, 59, 60, 61, 63, 66, 75, 79, 83, 93, 105
response time, 61
restrictions, 113
room temperature, 106
root, 63
rules, 86, 96

S

saturation, 63, 65, 79, 80, 82, 92, 93, 112
science, 118, 120, 124, 127
scope, 59, 100
security, 122
semiconductor, viii, 57, 59, 105, 106, 107, 108, 109, 110, 111, 112, 113, 114, 115, 116
sensitivity, 126
services, 58
shape, 63, 78, 83
signal processing, vii, viii, 57, 59, 60, 65, 72, 82, 105, 106, 107, 108, 109, 113, 114, 115, 116
signals, vii, 59, 61, 62, 63, 65, 67, 71, 73, 74, 75, 76, 79, 80, 81, 82, 84, 85, 86, 87, 89, 91, 92, 95, 103, 111
signal-to-noise ratio, 64, 112
silica, 61
simulation(s), viii, 58, 71, 77, 79, 80, 83, 86, 87, 95, 103, 110
Singapore, 117
smoothing, 63
society, 105
software, 120
solution, 61, 81
SS, 106
stability, 62
standard deviation, 93
state(s), viii, 58, 66, 68, 73, 74, 77, 79, 80, 82, 89, 98, 101
storage, 58, 126
structure, viii, 57, 65, 73, 77, 81, 104, 109
substrate, 66
succession, 64
Sun, 106, 108, 110, 113, 114, 116, 123
super-symmetry, vii, 1, 48, 49
suppression, 62, 112
susceptibility, 64
symmetry, vii
synchronization, 64
synchronize, 61
synthesis, 125
system-oriented levels, vii

T

target, viii, 58, 59, 67, 71, 76, 81, 88, 96, 104
techniques, vii, 127
technologies, viii, 58, 60, 105
technology, viii, 57, 79, 105, 110
temperature, 81, 85, 115
terminals, 92
textbooks, 127
thermal stability, 65
topology, 63
trade, 61, 64, 69
trade-off, 61, 64, 69
transmission, 60, 103
transparency, 61, 63, 108
treatment, 120
trial, 70

U

uniform, 60, 62, 69, 70, 95, 97

V

valence, 66, 67
variables, 126
variations, 77
vector, 117
velocity, 69, 86
vision, 105

W

wavelengths, 82, 85, 87, 89
wetting, 66, 68

Y

yield, 85